SPRINGER TRACTS IN MODERN PHYSICS

Ergebnisse
der exakten Natur-
wissenschaften

Volume 67

Springer-Verlag Berlin Heidelberg GmbH 1973

Manuscripts for publication should be adressed to:

G. HÖHLER, Institut für Theoretische Kernphysik der Universität, 75 Karlsruhe 1, Postfach 6380

Proofs and all correspondence concerning papers in the process of publication should be addressed to:

E. A. NIEKISCH, Kernforschungsanlage Jülich, Institut für Technische Physik, 517 Jülich, Postfach 365

ISBN 978-3-662-15570-7 ISBN 978-3-540-38498-4 (eBook)
DOI 10.1007/978-3-540-38498-4

Conformal Algebra in Space-Time and Operator Product Expansion*

S. FERRARA, R. GATTO, and A. F. GRILLO

Contents

---

* Dedicated to Prof. G. Höhler on his 50th birthday.

1. Introduction

The crucial role of short distance behaviour in elementary particle theory is widely recognized. In the last years the experiments at SLAC on deep inelastic electron scattering [1, 15, 26, 153, 130] have dramaticaly evidenced the fundamental role played by short-distance behaviour. In this case one is indeed exploring the behaviour at short distance of a product of two electromagnetic currents. Deep inelastic neutrino reactions [30, 125, 39, 134] are similarly related to the product of two weak currents. The forthcoming developments at NAL will provide additional stimulus and a new bunch of data on a wider range.

The short distance behaviour of an operator product of e.m. or weak currents is, of course, not only important for deep inelastic scattering, but also plays an important role in the theoretical description of other phenomena (such as for instance non-leptonic decays, electron pair annihilation, etc.). Wilson [159] has given a deep and original discussion of the operator product expansion and of its applications.

Wilson's expansion for the product of two currents has the form

$$j_\mu(x)\, j_\nu(y) = \sum_n c^n_{\mu\nu}(x-y)\, O_n(y)\,. \tag{1.1}$$

In Eq. (1.1) $c^n_{\mu\nu}(x-y)$ are c-numbers and $O_n(y)$ are local operators. The c-numbers are generally singular at short-distance. The expansion has been discussed in different theoretical frameworks, such as perturbation theory [20, 163, 164, 159, 165] and in model field theories. There has been particular discussion of operator expansion in the Thirring model [154, 103, 159, 160, 117, 50]. The conclusions strongly support the conjecture of the existence of an operator product expansion.

In his important paper [159] Wilson expecially emphasized the idea of using the concept of operator product expansion together with the assumption of scale invariance at short distances. Scale invariance has been invoked since many years in elementary particle theory [106, 108, 118, 119, 159].

In the expansion in Eq. (1.1) scale invariance allows for the determination of the degree of singularity of the coefficient $c^n_{\mu\nu}(x-y)$. The determination is based simply on comparison of the "scale dimensions" of the left- and right-hand-sides in Eq. (1.1). It is then of preliminary importance to know whether the operators O_n have definite "scale dimensions" or not. And, in case they have definite "scale dimensions", to know their actual values. Field theory investigations on this point have been very intricate, difficult, and not yet conclusive [31, 43, 160, 148–151, 44, 91, 161].

On the other hand the set of operators O_n is expected to contain at least some operator of well defined dimension, such as the currents (weak and electromagnetic, or, more generally, the unitary currents), and the stress-energy tensor. The currents have dimensions 3 in mass units and the stress-energy tensor has dimension 4. The existence (necessary in any local covariant theory) of a stress tensor $\theta_{\mu\nu}$ of definite dimension 4 has led to the suggestion that the scaling observed at SLAC is in fact explainable on the basis of the presence of $\theta_{\mu\nu}$ among the operators O_n in the right-hand-side of Eq. (1.1) [40]. The most simple explanation in terms of a single tensor of dimension 4, namely $\theta_{\mu\nu}$, seems in conflict with the different scaling of neutron and proton data. A more detailed argument making use of the universality of the coupling of $\theta_{\mu\nu}$ can however be developed and provides a more general justification [120] (a point which has particularly been emphasized by Wilson [162]).

Once scaling is accepted, as it must anyway at least approximately because of experiment, it can be shown by different arguments that a subset of operators O_n must exist, in Eq. (1.1), with dimensions d related to their spin by $d = s + 2$. Among them is the energy-momentum tensor with $s = 2$, and $d = 4$.

A simple way to demonstrate such a statement is from the convergence of the Bjorken-Callan-Gross sum rules [31, 13, 33]. Following the general treatment by Bjorken [13] one has the sum rules

$$\begin{aligned} f_n &= \int_0^2 \mathrm{d}\omega\, \omega^n F_t(\omega) \\ &= \lim_{p_z \to \infty} \frac{1}{2} (-i)^n \int \mathrm{d}^3 x \left\langle p_z \left| \left[\frac{\partial^n}{\partial t^n} j_x(x, 0), j_x(0, 0) \right] \right| p_z \right\rangle \Big/ p_0^n \end{aligned} \tag{1.2}$$

for each $n = 1, 3, 5, \ldots$ In Eq. (1.2) ω is the scaling variable, $\omega = -q^2/2\nu$, in the usual notations, and $F_t(\omega)$ is the transverse scaling function; j_x is the x-component of the e.m. current and $p = (p_0, 0, 0, p_z)$ is the nucleon momentum. The positivity of $F_t(\omega)$ ensures that each momentum integral f_n in Eq. (1.2) is positive and that the f_n's form a monotonic decreasing sequence. The relation $d = s + 2$ comes out in the following way. In Eq. (1.2) the matrix element of $[\partial^n j_x/\partial t^n, j_x]$ carries $n + 1$ momenta. The leading term in the commutator is then contributed by some operator, which we call $O_{\mu_1 \ldots \mu_{n+1}}$, which has spin $s = n + 1$ and dimensions, in mass units, $d = n + 3$. Thus $d = s + 2$, that is, one has a particular relation to satisfy between spin and dimension.

The relation would indeed easily be satisfied in a quark model through the (uncritical) assertion that each local operator keeps its "canonical" dimensions, as in free field theory, where such operators are constructed as Wick products of free fields. But the situation is not expected to be so simple in renormalized field theories [159]. The problem is indeed a complicated dynamical one [31, 43, 160, 148–151, 44, 91, 161].

From work of Callan, Coleman and Symanzik (see Refs. [31, 43, 148–151, 44]) it appears that, under some assumptions, one obtains renormalized dimensions (i.e. not necessarily canonical dimensions, but at least definite dimensions) in the so-called deep euclidean region and possibly also in the situation when the four-vector $x_\mu - y_\mu$ vanishes (the tip of the light-cone). The extension of such arguments to the whole light-cone is not yet known. For the operators relevant to an expansion on the whole light-cone [21, 22, 29, 95, 23, 71, 81, 82, 100, 114, 115, 25, 72, 167, 73, 147, 87, 45] one would have to prove much more, than just definite dimensions, namely, the relation $d = s + 2$ (canonical dimensions).

The argument showing that the scaling behaviour results from the configuration space behaviour on the light cone is well known (see for instance Ref. [95]). In the Fourier transform

$$\int e^{iqx} \langle p \,|\, [j_\mu(x), j_\nu(0)] \,|\, p\rangle \; d^4 x \qquad (1.3)$$

of the current correlation function one can write the exponent asymptotically as

$$qx = \nu x_0 - x_3(\nu + m/\omega)$$

by choosing the rest frame for the nucleon ($p = (m, 0)$) and the z direction along the vector $\boldsymbol{q}$. For $\nu \to \infty$ the integrand at $x_0 \approx x_3$ gives the dominant contribution, and because of the causal structure of the commutator this means that one is indeed exploring the light-cone. (For discussion on the relations between the $p \to \infty$ method with equal time commutators and the light cone, see for instance Refs. [113, 2, 102, 112, 45].) In fact one is exploring the whole light cone and not only its tip.

Note however that the positivity conditions play a fundamental role in that they allow to formulate restrictions on the operator product expansion, only by making use of its behaviour near the tip of the light-cone. This shows the power of the equal-time-commutator approach [33, 46, 13, 34, 47] and suggests the predominant role of the stress-energy tensor in the problem of scaling [40, 120].

When the naive quark commutators are assumed [73, 45, 87] both approaches, the one using equal time commutators or small distance behaviour of operator products, and the one postulating a quark-model light-cone expansion (possibly including gluon coupling [11, 24, 73, 87, 45]) reproduce a full set of testable relations [116, 121, 73, 45, 87, 49], which show the almost complete identity of such models to the more explicit parton model by Feynman [68, 69] and Bjorken [12, 14] (see Refs. [52–60, 83, 136, 137, 171]). In the parton model the virtual photon interacts very rapidly with the constituent as compared to the typical interaction time among constituents, which can therefore be assimilated to free particles. The parton model predictions are however more detailed in that they are based on particular parton distributions.

The approach using the algebraic concept of scale invariance was advocated by Wilson in connection to operator product expansion [159] and soon after adopted in the theory of deep electroproduction [40, 120, 79, 80, 152]. Previous work on the algebraic notion of dilatation invariance had been developed earlier on the basis of independent theoretical speculations [106, 108, 118, 119, 159].

The extension from dilatations to the entire conformal algebra (see Refs. 51, 89, 126, 127, 158, 74, 75, 104, 90, 105, 37, 107, 109, 38, 122, 32, 18, 88, 152, 35, 141]) may be justified on the basis of a number of reasons (none of them being however compelling):

(i) Lagrangian field theories which are formally invariant under dilatations are often invariant also under special conformal transformations. For instance, a sufficient condition (but not necessary) is the absence of derivative couplings.

(ii) Conformal transformations leave the light-cone invariant.

(iii) The conformal algebra provides for a natural homogeneization of the inhomogeneous Poincaré algebra.

The algebraic implications of conformal invariance on the light cone were studied, in Ref. [41], within the formalism using equal-time commutators. The requirement of covariance under the infinitesimal generators of $SU(2,2)$ (the covering group of the conformal group) can directly be imposed on an operator product expansion on the light-cone [63, 64, 66]. To such purpose one has first to analyze the transformation properties of the infinite set of local operators which provide a basis for the operator expansion [64].

It will be interesting to preliminarily illustrate two aspects of light-cone expansions which finally turn out to be (rather misteriously at first sight) connected to conformal invariance. They are: (a) causality, and, (b) translation invariance on a hermitean basis.

Let us consider the expansion

$$A(x)\,B(0) = \sum_n c_n(x)\,O_n(0) \tag{1.4}$$

where $c_n(x)$ are c-number functions and $O_n(x)$ form a complete set, extending the concepts and definitions of Wilson's work [159]. When we commute with some arbitrary local operators $C(y)$ we obtain

$$[A(x), C(y)]\,B(0) + A(x)\,[B(0), C(y)] = \sum_n c_n(x)\,[O_n(0), C(y)] \tag{1.5}$$

One notes that for y_μ spacelike, that is for $y^2 < 0$, each commutator $[O_n(0), C(y)]$ vanishes. So, taking $y^2 < 0$, each term on the right hand side vanishes, whereas on the left hand side $[B(0), C(y)]$ vanishes, but not necessarily does so the first term, $[A(x), C(y)]\,B(0)$. The latter term vanishes if in addition $(x-y)^2 < 0$, which, on the light cone amounts to requiring $y^2 < 2xy$. There is no paradox, because of the infinite summation on the right-hand-side. However it would be better to have an improved operator product expansion which formally exhibits the causality properties in each of its terms. This is the problem we have indicated under (a).

Let us now show what the problem specified under (b) is. Again let us consider a light cone expansion of the form

$$A(x)\,B(0)=(-x^2+i x_0)^{-\gamma}\sum_n x^{\alpha_1}\ldots x^{\alpha_n}\,O_{\alpha_1\ldots\alpha_n}(0). \tag{1.6}$$

The tensors $O_{\alpha_1\ldots\alpha_n}$ are symmetric traceless tensors. They can always be chosen to be hermitean [159, 25]. The commutator $[A(x), B(0)]$ has then the correct support required by causality. On Eq. (1.6) one makes a number of algebraic steps. One first translates by $-x$, then changes x into $-x$, expands $O_{\alpha_1\ldots\alpha_n}(x)$ in a power series around $x=0$, and finally takes the hermitean conjugate. The expression one finds can be compared with the expression one had started from, Eq. (1.6), and one discovers that to obtain consistency the following infinite set of relations has to be satisfied

$$O_{\alpha_1\ldots\alpha_n}(x)=\sum_{m=0}^{n}\frac{(-1)^m}{(n-m)!}\,\partial_{\alpha_{m+1}}\ldots\partial_{\alpha_n}O_{\alpha_1\ldots\alpha_n}(x). \tag{1.7}$$

It is interesting that, for $n=$ odd, Eq. (1.7) tells that $O_{\alpha_1\ldots\alpha_n}$ is a sum of derivative operators (which therefore have vanishing forward matrix elements).

It turns out that the imposition, to the general operator product expansion on the light-cone, of the requirement of covariance under infinitesimal special conformal transformations results in a very stringent set of limitations [63, 64, 66]. They essentially amount to fixing the relative coefficients in the expansion of each derivative term

$$\partial_{\alpha_{m+1}}\ldots\partial_{\alpha_n}O_{\alpha_1\ldots\alpha_m}(x)$$

with respect to the non-derivative term $O_{\alpha_1\ldots\alpha_m}(x)$. The expansion including such restrictions is found to exhibit a very compact form in terms of a confluent hypergeometric function. The interesting circumstance becomes then apparent, that the two problems we have mentioned under the headings (a) and (b) above, are in fact automatically solved with the new form of the operator product expansion, essentially reducing to some known properties of the confluent hypergeometric function. That is, the imposition of conformal invariance directly eliminates the two problems of causality support and translation invariance on a hermitean basis (the reverse however is not true).

We have already discussed why the imposition of conformal symmetry on the light cone seems to be a reasonable requirement. At this stage one is working within well-defined limitations: (i) one only deals with infinitesimal conformal transformations, and (ii) the symmetry is supposed to hold only on the light cone, but not necessarily for the complete theory. If conformal invariance is predominantly spontaneously broken, then the requirement of covariance of the operator product expansion under the conformal group may correspondingly enjoy of a larger domain of validity.

A most elegant derivation of a manifestly conformal covariant operator product expansion can be given by exploiting the isomorphism between the conformal algebra and the $O(4,2)$ orthogonal algebra. The derivation is uniquely and in a straightforward way extensible off the light-cone onto the entire space-time. Such expansions, manifestly conformal-covariant over the entire space-time, should apply to the skeleton theory, in Wilson's sense [159], provided it enjoys of the property of conformal invariance, beyond the postulated scale invariance. As we have said, simple Lagrangian theories which are invariant under dilatations turn out, under some general assumption to be also invariant under special conformal transformations; this may justify the hypothesis of a skeleton theory which is fully conformally invariant.

An important result appears to be a general theorem [63, 67], which we call the theorem of spin-dimension correlation, which exhibits the dynamical content of the relation $d = s + 2$, equivalent to the requirement of scaling in the Bjorken limit, in terms of a set of generalized partial conservation equations. By this we mean that the divergences of those tensors $O_{\alpha_1 \ldots \alpha_n}$, which contribute to the structure functions in the Bjorken limit, are annihilated by the generators of special conformal transformations K_λ. Of course, the special situation provided by the quark model commutators is a particular case of the above results, which can be regarded as a statement of the necessary and sufficient conditions for scaling (through the mechanism of canonical dimension for the relevant set of operators).

The manifestly covariant formulation in a six-dimensional coordinate space, based on the isomorphism with $O(4,2)$, is particularly useful in allowing for a direct and simple construction of vacuum expectation values of field products [67]. Among these, in particular, the three-point-functions, which are completely fixed (except from a constant) [128, 138, 145], are directly related to the covariant form of the operator product expansion. Precisely, the two problems, of constructing a conformally covariant operator product expansion, and of constructing the general covariant three-point-function are essentially equivalent formulations of the same problem [128, 65, 66]. This holds both on the light cone [128, 65] and off the light cone [66]. More generally, conformal covariance restricts the form of the n-point function, although only for $n = 3$ is the restriction capable of a unique prediction (apart from a constant). The simplest case, $n = 2$, i.e. the correlation function for two local operators, appears to be subject to very stringent limitations, which take the form of selection rules. That is, the two-point function vanishes unless the spins and dimensions are correlated in a precise way, if one assumes only conformal invariance on the light-cone; and it vanishes quite generally unless the operators have same spin and dimension, under the stronger assumption of full conformal covariance.

The implications of these results seem rather powerful, as they point out to strongly limited possibilities for a conformally invariant skeleton theory. In addition to these restrictions one has to recall that the causality limitations for the commutator of two local observables imposes rather strict constraints to a theory invariant under finite conformal transformations. For this problem however we refer to a comprehensive investigation by Kastrup et al. [110].

Furthermore, whenever gauge invariance constraints apply to conformaly covariant vacuum expectation values of products of local observables one generally obtains a stronger set of selection rules [128]. It therefore appears that much work has still to be carried out in order to develop a comprehensive understanding of the structure of a (broken) conformally invariant theory.

2. Introduction to the Conformal Group in Space-Time

2.1. The Conformal Group

The conformal group provides for an extension of the Poincaré group into a higher dimension homogeneous orthogonal group (see Refs. 51, 89, 126, 127, 133, 158, 92, 74, 75, 104–109, 90, 16, 37, 70, 38, 122, 123, 32, 88, 152, 35]). The conformal generators $J_{AB}(A, B=0, 1 \dots 6)$ satisfy the commutation relations

$$[J_{AB}, J_{CD}] = i(g_{AD}J_{BC} + g_{BC}J_{AD} - g_{AC}J_{BD} - g_{BD}J_{AC}) \tag{2.1}$$

where g_{AB} is diagonal with $g_{AA} = (+ - - -, - +)$. One has $J_{AB} = -J_{BA}$ giving a total of 15 independent generators. In terms of the $IO(3,1)$-generators $M_{\mu\nu}, P_\mu$, plus the new generators D and K_μ one has the correspondence

$$J_{AB} = \begin{array}{c|c|c|c|} & 0 \quad\quad 3 & 5 & 6 \\ \hline 0 \dots 3 & M_{\mu\nu} & & \\ \hline 5 & \tfrac{1}{2}(P_\mu - K_\mu) & & \\ \hline 6 & \tfrac{1}{2}(P_\mu + K_\mu) & & \end{array} \tag{2.2}$$

The algebra satisfied by the generators, Eq. (2.1), is the $O(4,2)$ algebra. It is isomorphic to the spinor algebra $SU(2,2)$. Written in terms of $M_{\mu\nu}$, P_μ, D, and K_μ the commutation relations in Eq. (2.1) become: Poincaré subalgebra

$$[M_{\mu\nu}, M_{\varrho\sigma}] = -i(g_{\mu\varrho}M_{\nu\sigma} - g_{\nu\varrho}M_{\mu\sigma} + g_{\mu\sigma}M_{\varrho\nu} - g_{\nu\sigma}M_{\varrho\mu}),$$
$$[M_{\mu\nu}, P_\varrho] = i(g_{\nu\varrho}P_\mu - g_{\mu\varrho}P_\nu) \qquad [P_\mu, P_\nu] = 0. \tag{2.3}$$

Lorentz behaviour of D (dilation generator)

$$[M_{\mu\nu}, D] = 0 \qquad \text{(i.e., } D \text{ is Lorentz scalar)}, \tag{2.4}$$

$$[P_\mu, D] = iP_\mu \qquad (P_\mu \text{ acts as a step-up operator with respect to } D). \tag{2.5}$$

Lorentz behaviour of K_μ (special conformal generator)

$$[M_{\mu\nu}, K_\varrho] = -i(g_{\varrho\mu}K_\nu - g_{\varrho\nu}K_\mu) \qquad (K_\mu \text{ is a four-vector}), \tag{2.6}$$

$$[P_\mu, K_\nu] = i2(g_{\mu\nu}D - M_{\mu\nu}). \tag{2.7}$$

$K_\mu - D$ commutators

$$[K_\mu, K_\nu] = 0, \tag{2.8}$$

$$[D, K_\mu] = iK_\mu \qquad (K_\mu \text{ acts as a step-down operator with respect to } D). \tag{2.9}$$

Linear realizations of the group of conformal transformations can be obtained in six-dimensions. The transformations are those which leave invariant a bilinear form with metric $(+---,-+)$. In four dimensions (Minkowski space) the realization is non-linear. One has:

$$x'_\mu = a_\mu + \Lambda_\mu^\nu x_\nu \qquad \text{(generators of infinitesimal transformations: } P_\mu, M_{\mu\nu}), \tag{2.10}$$

$$x'_\mu = e^\lambda x_\mu \qquad \lambda \text{ real (dilatations: generator } D), \tag{2.11}$$

$$x'_\mu = \frac{x_\mu + c_\mu x^2}{1 + 2c_\nu x^\nu + c^2 x^2}, \quad c_\mu \text{ real} \quad \text{(special conformal transformations: generators } K_\mu). \tag{2.12}$$

The independent parameters are: $4(a_\mu) + 6(\Lambda_\mu^\nu) + 1(\lambda) + 4(c_\mu) = 15$. Special conformal transformations can be thought as products of (inversion) $\times$ (translation) $\times$ (inversion):

$$x_\mu \to x_\mu/x^2 \to x_\mu/x^2 + c_\mu \to \frac{x_\mu/x^2 + c_\mu}{(x_\mu/x^2 + c_\mu)^2} = \frac{x_\mu + c_\mu x^2}{1 + 2cx + c^2x^2}$$

The group has two abelian subgroups: one generated by P_μ the other by K_μ. It has two Poincaré subalgebras:

$$(M_{\mu\nu}, P_\mu) \quad \text{and} \quad (M_{\mu\nu}, K_\mu) \tag{2.13}$$

2.2. Mass Spectrum

From $[P_\mu, D] = iP_\mu$ one has

$$[P^2, D] = P^\mu [P_\mu, D] + [P^\mu, D] P_\mu = 2iP^2 . \tag{2.14}$$

Suppose there is a discrete state of mass m whose normalized ket is $|p\rangle$, where $p^2 = m^2$. From Eq. (2.14)

$$\langle p|[P^2, D]|p\rangle = 2i\langle p|P^2|p\rangle = 2im^2 ,$$
$$\langle p|[P^2, D]|p\rangle = \langle p|m^2 D - Dm^2|p\rangle = 0 .$$

Thus: $m^2 = 0$. Discrete massive states are impossible, unless the symmetry is broken. Continuum massive states can however exist.

The argument here is similar to that of ordinary quantum mechanics showing that the spectrum of p is continuus from $[q, p] = i$. Again $\langle p'|[q,p]|p'\rangle = 0 = i\langle p'|p'\rangle$; however $\langle p'|qp - pq|p''\rangle = (p'' - p')\langle p'|q|p''\rangle = i\langle p'|p''\rangle = i\delta(p' - p'')$ showing that $q = i\,\partial/\partial p$. Indeed

$$(p'' - p')\, i(\partial/\partial p'')\, \langle p'|p''\rangle = i(p'' - p')\,(\partial/\partial p'')\,\delta(p' - p'') = i\delta(p' - p'') .$$

2.3. Representations

One uses the method of induced representations [124, 93, 101] (Mackey). Let us call G anyone of the generators of the stability subgroup at $x_\mu = 0 : G \equiv (M_{\mu\nu}, D, K_\mu)$. The remaining generator is P_μ, which acts like

$$[P_\mu, \varphi(x)] = -i\partial_\mu \varphi(x) . \tag{2.15}$$

To evaluate $[\varphi(x), G]$ we use Eq. (2.15)

$$\begin{aligned}[\varphi(x), G] &= [\mathrm{e}^{iPx}\varphi(0)\,\mathrm{e}^{-iPx}, G] = \mathrm{e}^{iPx}[\varphi(0), \tilde{G}]\,\mathrm{e}^{-iPx}\\ \tilde{G} &= \mathrm{e}^{-iPx} G\, \mathrm{e}^{iPx} .\end{aligned} \tag{2.16}$$

Let us calculate $\tilde{G}$. One has

$$\tilde{G} = \sum_n \frac{(-i)^n}{n!} x_{\mu_1} \dots x_{\mu_n} [P^{\mu_1}, [P^{\mu_2} \dots [P^{\mu_n}, G], \dots]] .$$

If $G \equiv M_{\mu\nu}$,

$$\begin{aligned}\tilde{M}_\mu &= M_\mu - ix_{\mu_1}[P^{\mu_1}, M_{\mu\nu}] - \tfrac{1}{2}x_{\mu_1}x_{\mu_2}[P^{\mu_1}, [P^{\mu_2}, M_{\mu\nu}],] + \cdots\\ &= M_{\mu\nu} - x_{\mu_1}(g_\nu^{\mu_1} P_\mu - g_\mu^{\mu_1} P_\nu) = M_{\mu\nu} - (x_\nu P_\mu - x_\mu P_\nu) .\end{aligned} \tag{2.17}$$

If $G \equiv D$

$$\tilde{D} = D - ix_{\mu_1}[P^{\mu_1}, D] - \tfrac{1}{2}x_{\mu_1}x_{\mu_2}[P^{\mu_1}[P^{\mu_2}, D],] + \cdots = D + x_\lambda P^\lambda . \tag{2.18}$$

If $G \equiv K_\mu$

$$\tilde{K}_\mu = K_\mu - ix_{\mu_1}[P^{\mu_1}, K_\mu] - \tfrac{1}{2}x_{\mu_1}x_{\mu_2}[P^{\mu_1}, [P^{\mu_2}, K_\mu],] + \cdots .$$

One has

$$[P^{\mu_1}, [P^{\mu_2}, [P^{\mu_3}, K_\mu],],] = 2i[P^{\mu_1}, [P^{\mu_2}, (g_\mu^{\mu_3} D - M_\mu^{\mu_3})],] = 0 .$$

Thus:

$$\begin{aligned}\tilde{K}_\mu &= K_\mu - ix_{\mu_1} 2i(g_\mu^{\mu_1} D - M_\mu^{\mu_1}) - \tfrac{1}{2} x_{\mu_1} x_{\mu_2} [P^{\mu_1}, 2i(g_\mu^{\mu_2} D - M_\mu^{\mu_2})] \\ &= K_\mu + 2x^\nu (g_{\mu\nu} D + M_{\nu\mu}) + 2x_\mu x^\nu P_\nu - x^2 P_\mu .\end{aligned} \tag{2.19}$$

In conclusion $\tilde{G}$ is always a finite linear combination of generators [135].

Let us now examine the representations of the stability subgroup. Within a representation that behaves irreducibly under $M_{\mu\nu}$, clearly D is a multiple of $\mathbb{1}$ (from Eq. (2.4)). Eq. (2.9) then tells that K_μ vanishes within the representation. We thus have

$$[\varphi(0), M_{\mu\nu}] = \Sigma_{\mu\nu} \varphi(0) , \tag{2.20}$$

$$[\varphi(0), D] = i\varphi \Delta(0) , \quad \text{where } \Delta \text{ is a } c\text{-number to be called scale dimension} \tag{2.21}$$

$$[\varphi(0), K_\mu] = 0 . \tag{2.22}$$

From Eqs. (2.16)–(2.18) and the above equations

$$\begin{aligned}[\varphi(x), M_{\mu\nu}] &= e^{iP\cdot x} [\varphi(0), M_{\mu\nu}] e^{-iP\cdot x} = e^{iP\cdot x} [\varphi(0), M_{\mu\nu} \\ &\quad - (x_\nu P_\mu - x_\mu P_\nu)] e^{-iP\cdot x} = \Sigma_{\mu\nu} \varphi(x) - i(x_\nu \partial_\mu - x_\mu \partial_\nu) \varphi(x) ,\\ [\varphi(x), D] &= e^{iP\cdot x} [\varphi(0), D] e^{-iP\cdot x} = e^{iP\cdot x} [\varphi(0), (D + x_\lambda P^\lambda)] e^{-iP\cdot x} \\ &= i\Delta \varphi(x) + ix_\lambda \partial^\lambda \varphi(x) ,\\ [\varphi(x), K_\mu] &= e^{iP\cdot x} [\varphi(0), \tilde{K}_\mu] e^{-iP\cdot x} \\ &= e^{iP\cdot x} [\varphi(0), K_\mu + 2x^\nu (g_{\mu\nu} D + M_{\mu\nu}) + 2x_\mu x^\nu P_\nu - x^2 P_\mu] e^{-iP\cdot x} \\ &= [2x^\nu (g_{\mu\nu} i\Delta + \Sigma_{\mu\nu}) + 2x_\mu x^\nu i\partial_\nu - ix^2 \partial_\mu] \varphi(x) .\end{aligned}$$

In conclusion

$$[\varphi(x), M_{\mu\nu}] = i[(x_\mu \partial_\nu - x_\nu \partial_\mu) - i\Sigma_{\mu\nu}] \varphi(x) , \tag{2.23}$$

$$[\varphi(x), D] = i(l + x^\varrho \partial_\varrho) \varphi(x) , \tag{2.24}$$

$$[\varphi(x), K_\mu] = i[(2\Delta x_\mu + 2x_\mu x^\varrho \partial_\varrho - x^2 \partial_\mu) + 2ix^\varrho \Sigma_{\mu\varrho}] \varphi(x) \tag{2.25}$$

and of course

$$[\varphi(x), P_\mu] = i\partial_\mu \varphi(x) . \tag{2.26}$$

Representations with non-vanishing K_μ are of two types: with K_μ represented by a nihilpotent matrix, or infinite dimensional.

Finite dimensional representations that behave reducibly under $M_{\mu\nu}$ can be analysed as before. $M_{\mu\nu}$ can be taken block-diagonal, in a suitable basis. The D is correspondingly block diagonal by the same argument used before, and again K_μ vanishes in each block. Since

$$[D, K_\mu^p] = ipK_\mu^p, \tag{2.27}$$

where p is any power also K_μ^p vanishes in each block. Eq. (2.27) follows by iterating Eq. (2.9). Then K_μ must be nihilpotent. Such a result is also intuitive since K_μ lowers the dimension and the representation is finite dimensional. Finite dimensional representations of the stability subgroup can therefore always be brought into block diagonal form for $M_{\mu\nu}$ and D with K_μ represented by a lower-diagonal matrix. Eq. (2.25) is then modified by the addition within the square bracket of the nihilpotent matrix K_μ. One can always assume that the eigenvalues of D have been partially ordered into increasing dimensions $\Delta_1, \Delta_2, \ldots$. One then sees that the representation contains a finite number of invariant subspaces each obtained by keeping all those states with dimension lower than some Δ_i. However the representation is not reducible within such invariant subspaces, as K_λ does not correspondly assume a diagonal form. Thus one has at the same time the existence of an ordered set of invariant subspaces but not full reducibility of the representation. Such representations are thus called indecomposible and they clearly can be extended to infinite dimensions [78, 101, 121, 63].

The integration of Eqs. (2.23) and (2.26) for finite transformations is well-known. Eqs. (2.24) and (2.25) become, in the integrated form (see for instance Ref. [98]).

$$\varphi'_\alpha(x') = e^{l\lambda}\varphi_\alpha(x)\,; \qquad x'_\mu = e^\lambda x_\mu\,,$$

$$\varphi'_\alpha(x') = \sigma(x,c)^{-l} S_\alpha^\beta(x,c)\,\varphi_\beta(x)\,; \qquad x'_\mu = \frac{x_\mu + c_\mu x^2}{1 + 2c\cdot x + c^2x^2}$$

and $\sigma(x,c) = 1 + 2c\cdot x + c^2x^2$.

For tensor fields of order n the matrices $S_\alpha^\beta(x,c)$ turn out to be:

$$S_{\alpha_1\ldots\alpha_n}^{\beta_1\ldots\beta_n}(x,c) = \sigma^{-n}(x,c)\,(\partial x^{\beta_1}/\partial x'^{\alpha_1} \ldots \partial x^{\beta_n}/\partial x'^{\alpha_n})\,.$$

2.4. Conformal Currents

2.4.1. Under an infinitesimal conformal transformation specified by the parameters $\delta\omega_A$ the variation of the action integral

$$I(\Omega) = \int_\Omega d^4x \mathscr{L}(x)$$

for an arbitrary space-time volume Ω takes on the form

$$\delta I(\Omega) = \sum_A \delta\omega_A \int d^4x\, \partial_\mu J^{A\mu}\,. \tag{2.28}$$

For exact symmetry $\delta I(\Omega)=0$ and the currents J_μ^A are divergenceless.

Noether's theorem provides for canonical currents satisfying Eq. (2.28). They are defined from

$$\sum_A J_\mu^A \delta\omega_A = \pi_\mu \delta\varphi - \mathcal{T}_{\mu\nu}\delta x^\nu . \tag{2.29}$$

In Eq. (2.29) we have omitted for brevity summation over the independent fields: $\pi_\mu \delta\varphi$ stands for $\pi_\mu^a \delta\varphi_a$ where a runs over the different fields; $\mathcal{T}_{\mu\nu}$ is the canonical energy-momentum tensor

$$\mathcal{T}_{\mu\nu} = -g_{\mu\nu}\mathcal{L} + \pi_\mu \partial_\nu \varphi \tag{2.30}$$

and

$$\delta\varphi = \varphi'(x') - \varphi(x) , \tag{2.31}$$

$$\delta x_\nu = x'_\nu - x_\nu , \tag{2.32}$$

$$\pi_\mu = \partial\mathcal{L}/\partial(\partial_\mu\varphi) . \tag{2.33}$$

Equivalently one can write

$$\sum_A J_\mu^A \delta\omega_A = \pi_\mu \delta\varphi + \mathcal{L}\delta x_\mu - \pi_\mu(\partial_\nu\varphi)\,\delta x^\nu = \mathcal{L}\delta x_\mu + \pi_\mu \Delta\varphi , \tag{2.34}$$

$$\begin{aligned} \Delta\varphi &= \delta\varphi - (\partial_\nu\varphi)\,\delta x^\nu = \varphi'(x') - \varphi(x) - [\varphi(x') - \varphi(x)] \\ &= \varphi'(x') - \varphi(x') \cong \varphi'(x) - \varphi(x) . \end{aligned} \tag{2.35}$$

From Eq. (2.29), inserting Eqs. (2.10)–(2.12) and (2.23)–(2.26) one obtains the canonical conformal currents (see e.g. Ref. [122]):

$$\mathcal{T}_{\mu\nu} = -g_{\mu\nu}\mathcal{L} + \pi_\mu \partial_\nu \varphi \qquad \text{(translations)} , \tag{2.36}$$

$$\mathcal{M}^{(c)}_{\varrho\mu\nu} = x_\mu \mathcal{T}_{\varrho\nu} - x_\nu \mathcal{T}_{\varrho\mu} - i\pi_\varrho \Sigma_{\mu\nu}\varphi \qquad \text{(homogeneous Lorentz transformations)} , \tag{2.37}$$

$$\mathcal{D}^{(c)}_\mu = x^\nu \mathcal{T}_{\mu\nu} - \pi_\mu L\varphi \qquad \text{(dilatations)} , \tag{2.38}$$

$$\mathcal{K}^{(c)}_{\mu\nu} = 2x_\nu x^\varrho \mathcal{T}_{\mu\varrho} - x^2 \mathcal{T}_{\mu\nu} - \pi_\mu [2Lx_\nu + 2ix^\varrho \Sigma_{\varrho\nu}]\,\varphi \qquad \text{(special conformal transformations)} . \tag{2.39}$$

In Eqs. (2.38) and (2.39) L is the dimension matrix, taken to be diagonal, with eigenvalue $-\Delta^a$ for a-th field.

2.4.2. As well-known $\partial^\mu \mathcal{T}_{\mu\nu}=0$ is equivalent to the field equations:

$$\partial^\mu \mathcal{T}_{\mu\nu} = -\partial_\nu \mathcal{L} + \partial^\mu(\pi_\mu \partial_\nu\varphi) = -(\partial_\nu\varphi)\left[\frac{\partial\mathcal{L}}{\partial\varphi} + \partial^\mu \pi_\mu\right] = 0 \tag{2.40}$$

and $\partial^\varrho \mathcal{M}^{(c)}_{\varrho\mu\nu}$ is equivalent to Lorentz invariance of the Lagrangian

$$\partial^\varrho \mathcal{M}^{(c)}_{\varrho\mu\nu} = \mathcal{T}_{\mu\nu} - \mathcal{T}_{\nu\mu} - \partial^\varrho(i\pi_\varrho \Sigma_{\mu\nu}\varphi) = 0 . \tag{2.41}$$

The tensor $\mathcal{T}_{\mu\nu}$ is symmetric only for spinless fields, $\Sigma_{\mu\nu}=0$. One has similarly

$$\partial^\mu \mathcal{D}_\mu^{(c)} = \mathcal{T}_\nu^{\ \nu} - \partial^\mu(\pi_\mu L\varphi) \tag{2.42}$$

and

$$\partial^\mu \mathcal{K}_{\mu\nu}^{(c)} = 2x_\nu \partial^\mu \mathcal{D}_\mu^{(c)} + \mathcal{R}_\nu \tag{2.43}$$

where

$$\mathcal{R}_\nu = -2\pi_\nu L + 2i\pi^\mu \Sigma_{\mu\nu}\varphi\,. \tag{2.44}$$

Let us consider a simplest example: a massless spinless field. One has

$$\mathcal{L} = -\tfrac{1}{2}(\partial_\mu\varphi)(\partial^\mu\varphi)\,, \tag{2.45}$$

$$\pi_\mu = -\partial_\mu\varphi\,, \tag{2.46}$$

$$\mathcal{T}_{\mu\nu} = \tfrac{1}{2}g_{\mu\nu}(\partial_\lambda\varphi)(\partial^\lambda\varphi) - (\partial_\mu\varphi)(\partial_\nu\varphi)\,, \tag{2.47}$$

from which

$$\mathcal{T}_\nu^{\ \nu} = (\partial_\lambda\varphi)(\partial^\lambda\varphi)\,, \tag{2.48}$$

$$\partial^\mu \mathcal{D}_\mu^{(c)} = (\partial_\lambda\varphi)(\partial^\lambda\varphi) + \partial^\mu(\partial_\mu\varphi)L\varphi = (\partial_\lambda\varphi)(1+L)(\partial^\lambda\varphi) \tag{2.49}$$

which indeed vanishes for the canonical value $L=-1$.

For $\mathcal{K}_{\mu\nu}^{(c)}$ one has

$$\partial^\mu \mathcal{K}_{\mu\nu}^{(c)} = \mathcal{R}_\nu = 2(\partial_\nu\varphi)L\varphi = -\partial_\nu\varphi^2\,. \tag{2.50}$$

Eq. (2.50) shows that the theory is apparently not conformally invariant. But this is only due to the particular choice of $\mathcal{K}_{\mu\nu}^{(c)}$. We can always add to $\mathcal{K}_{\mu\nu}^{(c)}$ a term $g_{\mu\nu}\varphi^2$ which does not modify the commutators of the conformal charges with the fields and adds to $\partial^\mu \mathcal{K}_{\mu\nu}^{(c)}$ only a divergence, $\partial_\nu\varphi^2$. Therefore in general (we assume for simplicity that only fields of spin $\leqq 1$ are present) we shall redefine $\mathcal{K}_{\mu\nu}^{(c)}$ in the form

$$\mathcal{K}_{\mu\nu}^{(c)} = 2x_\nu x^\varrho \mathcal{T}_{\mu\varrho} - x^2\mathcal{T}_{\mu\nu} - \pi_\mu[2Lx_\nu + 2ix^\varrho\Sigma_{\varrho\nu}]\varphi + g_{\mu\nu}\sum_i \varphi_i^+\varphi_i \tag{2.51}$$

where φ_i are the spinless fields of the theory.

Eq. (2.43) now becomes

$$\partial^\mu \mathcal{K}_{\mu\nu}^{(c)} = 2x_\nu \partial^\mu \mathcal{D}_\mu^{(c)} + \mathcal{R}_\nu\,, \tag{2.52}$$

$$\mathcal{R}_\nu = -2\pi_\nu L\varphi + 2i\pi^\mu\Sigma_{\mu\nu}\varphi + \partial_\nu\sum_i \varphi_i^+\varphi_i \tag{2.53}$$

$\mathcal{R}_\nu$, as defined in Eq. (2.53), now vanishes for canonical kinetic energy terms. Thus

$$\mathcal{R}_\nu = -2\frac{\partial\mathcal{L}'}{\partial(\partial_\nu\varphi)}L\varphi + 2i\frac{\partial\mathcal{L}'}{\partial(\partial_\mu\varphi)}\Sigma_{\mu\nu}\varphi \tag{2.54}$$

where $\mathscr{L}'$ is the rest of the lagrangian (that is excluding the kinetic energy terms of canonical form). For theories where $\mathscr{L}'$ does not contain derivatives one has $\mathscr{R}_\nu = 0$ and Eq. (2.52) tells that scale invariance, $\partial^\mu \mathscr{D}_\mu^{(c)} = 0$, implies the entire conformal invariance $\partial^\mu \mathscr{K}_{\mu\nu}^{(c)} = 0$. The known renormalizable field theory models are of this kind.

The relation between scale invariance and absence of dimensioned constants can be understood in the following way. Eq. (2.42) tells us that

$$\begin{aligned}\partial^\mu \mathscr{D}_\mu^{(c)} &= -4\mathscr{L} + \pi_\lambda \partial^\lambda \varphi - \pi_\lambda L \partial^\lambda \varphi - (\partial \mathscr{L}/\partial \varphi) L\varphi \\ &= -4\mathscr{L} + \left[(-\partial \mathscr{L}/\partial \varphi) L\varphi + (1-L)\frac{\partial \mathscr{L}}{\partial(\partial_\lambda \varphi)}(\partial^\lambda \varphi)\right].\end{aligned} \tag{2.55}$$

Now if φ has dimensions (in energy units) $-L$, $\partial^\lambda \varphi$ has dimensions $-L+1$ and the term in bracket amounts just to $4\mathscr{L}$, by Euler's theorem, if $\mathscr{L}$ is indeed a homogeneous function of degree 4 in φ and $\partial^\lambda \varphi$.

In the following we shall consider a more general form of the condition $\mathscr{R}_\nu = 0$, required to have conformal invariance following from scale invariance. We shall only require that $\mathscr{R}_\nu = \partial_\nu F(\varphi^a)$ where $F(\varphi^a)$ depends on the fields in the theory [88]. If this conditions is verified one can again redifine $\mathscr{K}_{\mu\nu}^{(c)}$ as $\mathscr{K}_{\mu\nu}^{(c)} - g_{\mu\nu} F$. Eq. (2.32) becomes

$$\partial^\mu \mathscr{K}_{\mu\nu}^{(c)} = 2x_\nu \partial^\mu \mathscr{D}_\mu^{(c)} + \partial_\nu F \tag{2.56}$$

so that again

$$\partial^\mu (\mathscr{K}_{\mu\nu}^{(c)} - g_{\mu\nu} F) = 2x_\nu \partial^\mu \mathscr{D}_\mu^{(c)}. \tag{2.57}$$

2.4.3. We shall assume here that the condition $\mathscr{R}_\nu = \partial_\nu F$ holds and proceed to show that the conformal currents can be redefined as: $\theta_{\mu\nu}$, a symmetric energy momentum tensor, in place of $\mathscr{T}_{\mu\nu}$, $\mathscr{M}_{\varrho\mu\nu} = x_\nu \theta_{\mu\varrho} - x_\varrho \theta_{\mu\nu}$, $\mathscr{D}_\mu = x^\nu \theta_{\mu\nu}$, and $\mathscr{K}_{\mu\nu} = x_\nu \mathscr{D}_\mu + x^\varrho \mathscr{M}_{\mu\nu\varrho}$. In a conformally invariant theory such currents are conserved, and give rise to the same charges P_μ, $M_{\mu\nu}$, D, and K_μ. The result extends the well-known result of Belinfante [9, 10] and Møller [13] for the Lorentz subgroup. For a more general discussion see Callan, Coleman and Jackiw [32]. We sketch here the proof. Let us define:

$$H_{\varrho[\mu\nu]} = -i\pi_\varrho \Sigma_{\mu\nu} \varphi + \tfrac{1}{6}(g_{\varrho\mu}\partial_\nu - g_{\varrho\nu}\partial_\mu)(\varphi^2 + F(\varphi)) = -H_{\varrho[\nu\mu]}, \tag{2.58}$$

$$\mathscr{G}_{[\varrho\mu]\nu} = \tfrac{1}{2}(H_{\varrho[\mu\nu]} + H_{\mu[\nu\varrho]} + H_{\nu[\mu\varrho]}). \tag{2.59}$$

One has

$$\mathscr{G}_{[\mu\varrho]\nu} = \tfrac{1}{2}(H_{\mu[\varrho\nu]} + H_{\varrho[\nu\mu]} + H_{\nu[\varrho\mu]}) = -\mathscr{G}_{[\varrho\mu]\nu}. \tag{2.60}$$

We now define

$$\theta_{\mu\nu} = \mathscr{T}_{\mu\nu} + \partial^\varrho \mathscr{G}_{[\varrho\mu]\nu}. \tag{2.61}$$

From Eqs. (2.41) and (2.58)

$$\mathcal{T}_{\mu\nu} - \mathcal{T}_{\nu\mu} = \partial^\varrho (i\pi_\varrho \Sigma_{\mu\nu}\varphi) = -\partial^\varrho H_{\varrho[\mu\nu]} . \tag{2.62}$$

Now from Eqs. (2.61) and (2.59)

$$\begin{aligned} \theta_{\mu\nu} - \theta_{\nu\mu} &= \mathcal{T}_{\mu\nu} - \mathcal{T}_{\nu\mu} + \partial^\varrho (\mathcal{G}_{[\varrho\mu]\nu} - \mathcal{G}_{[\varrho\nu]\mu}) \\ &= \mathcal{T}_{\mu\nu} - \mathcal{T}_{\nu\mu} + \partial^\varrho H_{\varrho[\mu\nu]} = 0 \end{aligned} \tag{2.63}$$

because of Eq. (2.62).

Furthermore Eqs. (2.61) and (2.40) tell us that

$$\partial^\mu \theta_{\mu\nu} = \partial^\mu \partial^\varrho \mathcal{G}_{[\varrho\mu]\nu} = 0 \tag{2.64}$$

and

$$\int d^3x \theta_{0\mu} = P_\mu + \int d^3x \partial^\varrho \mathcal{G}_{[\varrho 0]\mu} = P_\mu + \int d^3x \partial^i \mathcal{G}_{[i0]\mu} = P_\mu . \tag{2.65}$$

This well-known proof has been given here in some detail because the other proofs for $\mathcal{M}_{\varrho\mu\nu} = x_\nu \theta_{\mu\varrho} - x_\varrho \theta_{\mu\nu}$, $\mathcal{D}_\mu = x^\nu \theta_{\mu\nu}$, and $\mathcal{K}_{\mu\nu} = x_\nu \mathcal{D}_\mu + x^\varrho \mathcal{M}_{\mu\nu\varrho}$ follow identical lines. One has, briefly, for $\mathcal{M}_{\mu\nu\varrho}$

$$\begin{aligned} \mathcal{M}_{\mu\nu\varrho} &= x_\nu \theta_{\mu\varrho} - x_\varrho \theta_{\mu\nu} = x_\nu \mathcal{T}_{\mu\varrho} - x_\varrho \mathcal{T}_{\mu\nu} + \partial^\sigma (x_\nu \mathcal{G}_{[\sigma\mu]\varrho} - x_\varrho \mathcal{G}_{[\sigma\mu]\nu}) \\ &\quad - (\mathcal{G}_{[\nu\mu]\varrho} - \mathcal{G}_{[\varrho\mu]\nu}) \\ &= x_\nu \mathcal{T}_{\mu\varrho} - x_\varrho \mathcal{T}_{\mu\nu} + i\pi_\mu \Sigma_{\varrho\nu}\varphi - \tfrac{1}{6}(g_{\mu\varrho}\partial_\nu - g_{\mu\nu}\partial_\varrho)(\varphi^2 + F(\varphi)) \\ &\quad - \partial^\sigma (x_\nu \mathcal{G}_{[\sigma\mu]\varrho} - x_\varrho \mathcal{G}_{[\sigma\mu]\nu}) , \end{aligned} \tag{2.66}$$

$$\begin{aligned} &\mathcal{M}^{(c)}_{\mu\nu\varrho} + \partial^\sigma (x_\nu \mathcal{G}_{[\sigma\mu]\varrho} - x_\varrho \mathcal{G}_{[\sigma\mu]\nu}) - \tfrac{1}{6}(g_{\mu\varrho}\partial_\nu - g_{\mu\nu}\partial_\varrho)(\varphi^2 + F(\varphi)) \\ &\quad = \mathcal{M}^{(c)}_{\mu\nu\varrho} - \partial^\sigma X_{[\sigma\mu]\nu\varrho} , \end{aligned} \tag{2.67}$$

$$X_{[\sigma\mu]\nu\varrho} = x_\varrho \mathcal{G}_{[\sigma\mu]\nu} - x_\nu \mathcal{G}_{[\sigma\mu]\varrho} + \tfrac{1}{6}(g_{\mu\varrho} g_{\sigma\nu} - g_{\mu\nu} g_{\sigma\varrho})(\varphi^2 + F(\varphi)) . \tag{2.68}$$

It follows that

$$\partial^\mu \mathcal{M}_{\mu\nu\varrho} = 0 \tag{2.69}$$

since $\partial^\mu \partial^\sigma X_{[\sigma\mu]\nu\varrho} = 0$, and also that

$$M_{\nu\varrho} = \int d^3x \mathcal{M}_{0\nu\varrho} = \int d^3x \mathcal{M}^{(c)}_{0\nu\varrho} - \int d^3x \partial^i X_{[i0]\nu\varrho} = \int d^3x \mathcal{M}^{(c)}_{0\nu\varrho} . \tag{2.70}$$

For $\mathcal{D}_\mu = x^\nu \theta_{\mu\nu}$ one similarly obtains

$$\begin{aligned} \mathcal{D}_\mu &= x^\nu \mathcal{T}_{\mu\nu} + \partial^\varrho (x^\nu \mathcal{G}_{[\varrho\mu]\nu}) - \mathcal{G}^\nu_{[\nu\mu]} = x^\nu \mathcal{T}_{\mu\nu} + i\pi^\nu \Sigma_{\mu\nu}\varphi \\ &\quad + \tfrac{1}{2}\partial_\mu(\varphi^2 + F(\varphi)) + \partial^\varrho (x^\nu \mathcal{G}_{[\varrho\mu]\nu}) = \mathcal{D}^{(c)}_\mu + \partial^\varrho (x^\nu \mathcal{G}_{[\varrho\mu]\nu}) \end{aligned} \tag{2.71}$$

after having inserted the relation

$$V_\mu = \partial_\mu F = -2\pi_\mu L\varphi + 2i\pi^\nu \Sigma_{\nu\mu}\varphi + \partial_\mu \varphi^2 .$$

Thus

$$\partial_\mu \mathcal{D}^\mu = \partial_\mu \mathcal{D}^{(c)\mu} \qquad D = \int d^3x \mathcal{D}_0 . \tag{2.72}$$

Quite similarly for $\mathscr{K}_{\mu\nu}$, after a slightly lengthier calculation, one finds

$$\mathscr{K}_{\mu\nu} = \mathscr{K}^{(c)}_{\mu\nu} + \partial^\sigma Y_{[\sigma\mu]\nu} \tag{2.73}$$

where

$$Y_{[\sigma\mu]\nu} = 2x^\varrho x_\nu \mathscr{G}_{[\sigma\mu]\varrho} - x^2 \mathscr{G}_{[\sigma\mu]\varrho} - \tfrac{1}{3}(x_\mu g_{\nu\sigma} - g_{\mu\nu} x_\varrho)(\varphi^2 + F(\varphi)). \tag{2.74}$$

The conclusion that has been reached is thus that the following charges can be defined in terms of the stress-energy tensor $\theta_{\mu\nu}$

$$P_\mu = \int d^3x \theta_{0\mu} \qquad M_{\mu\nu} = \int d^3x (x_\mu \theta_{0\nu} - x_\nu \theta_{0\mu}), \tag{2.75}$$

$$D = \int d^3x x^\mu \theta_{0\mu} \qquad K_\mu = \int d^3x (2x_\mu x^\sigma \theta_{0\sigma} - x^2 \theta_{0\mu}). \tag{2.76}$$

In a lagrangian field theory $\theta_{\mu\nu}$ is defined from Eq. (2.61)

$$\theta_{\mu\nu} = \mathscr{T}_\mu + \partial^\varrho \mathscr{G}_{[\varrho\mu]} = \theta_{\nu\mu} \tag{2.61}$$

where

$$\mathscr{T}_{\mu\nu} = -g_{\mu\nu}\mathscr{L} + \pi_\mu \partial_\nu \varphi, \tag{2.36}$$

$$\mathscr{G}_{[\varrho\mu]\nu} = \tfrac{1}{2}(H_{\varrho[\mu\nu]} + H_{\mu[\nu\varrho]} + H_{\nu[\mu\varrho]}), \tag{2.59}$$

$$H_{\varrho[\mu\nu]} = -i\pi_\varrho \Sigma_{\mu\nu}\varphi + \tfrac{1}{6}(g_{\varrho\mu}\partial_\nu - g_{\varrho\nu}\partial_\mu)(\varphi^2 + F(\varphi)). \tag{2.58}$$

When θ^μ_μ vanishes all conformal currents

$$\theta_{\mu\nu}, \mathscr{M}_{\varrho\mu\nu} = x_\nu \theta_{\mu\varrho} - x_\varrho \theta_{\mu\nu}, \mathscr{D}_\mu = x^\nu \theta_{\mu\nu}, \mathscr{K}_{\mu\nu} = x_\nu \mathscr{D}_\mu + x^\varrho \mathscr{M}_{\mu\nu\varrho}$$

are conserved (i.e. $\partial^\mu \theta_{\mu\nu} = 0$, $\partial^\mu \mathscr{M}_{\varrho\mu\nu} = 0$ always hold, and furthermore $\partial^\mu \mathscr{D}_\mu = 0$, $\partial^\mu \mathscr{K}_{\mu\nu} = 0$) as can directly be seen by inspection from Eq. (2.59). The charges in Eq. (2.59) then satisfy the algebra of the conformal charges in Eqs. (2.3)–(2.9).

3. Broken Conformal Symmetry

3.1. Broken Conformal Symmetry and the Energy Momentum Tensor

In the presence of symmetry breaking the algebra satisfied by the conformal charges

$$P_\mu(t) = \int d^3x \theta_{0\mu}(x) \qquad M_{\mu\nu}(t) = \int d^3x (x_\mu \theta_{0\nu}(x) - x_\nu \theta_{0\mu}(x)) \tag{3.1}$$

$$D(t) = \int d^3x x^\lambda \theta_{0\lambda}(x) \qquad K_\mu(t) = \int d^3x (2x_\mu x^\lambda - g^\lambda_\mu x^2)\theta_{0\lambda} \tag{3.2}$$

is no longer the conformal algebra. We shall be interested in finding out such algebra under some specific assumption on the nature of the symmetry breaking. In Eqs. (3.1) and (3.2) $\theta_{\mu\nu}$ is the energy-momentum tensor defined in such a way that P_μ and $M_{\mu\nu}$ in Eq. (3.1) are the

Poincaré generators. Such definition is of course not unique – a well-known circumstance. To further restrict the class of possible energy-momentum tensors we add the requirement that in the limit when $\theta(x)=\theta^{\lambda}_{\lambda}(x)$ vanishes one has exact conformal symmetry. The problem of investigating how $\theta_{\mu\nu}(x)$ is defined in a broken symmetry is of principal interest, and will be investigated first in the following sections.

We shall assume that $\theta(x)$ can be taken [80] as a sum of local scalars $u_j(x)$ which satisfy the equations

$$[u_j(0), D(0)] = i\Delta_j u_j(0)\,, \tag{3.3}$$

$$[u_j(0), K_\mu(0)] = 0\,. \tag{3.3'}$$

Eqs. (3.3) tell us that $u_j(x)$ transforms as a finite dimensional representation of the stability subgroup generated by $M_{\mu\nu}$, D, and K_μ, and has scale dimension Δ_j.

As we have just said, the first problem will be to investigate the definition of $\theta_{\mu\nu}$ in a broken symmetry. We shall follow here the treatment by Sartori and one of the authors [76]. One first notices that one can arbitrarily add to $\theta_{\mu\nu}$, defined such that P_μ and $M_{\mu\nu}$ in Eq. (3.1) are the Poincaré generators, a tensor of the form $\partial^\varrho G_{[\varrho,\mu]\nu}$ such that, – besides being antisymmetric in $\varrho\cdot\mu$ as indicated by the brackets, – one has

$$\partial^\varrho \mathscr{G}_{[\varrho,\mu]\nu} = \partial^\varrho \mathscr{G}_{[\varrho,\nu]\mu} \tag{3.4}$$

and

$$\mathscr{G}_{[\varrho,\mu]\nu} - \mathscr{G}_{[\varrho,\nu]\mu} = \partial^\sigma F_{[\sigma,\varrho][\mu,\nu]}\,. \tag{3.5}$$

Furthermore we can consistently require that

$$\partial^\varrho \mathscr{G}^\mu_{[\varrho,\mu]} \to 0 \quad \text{when} \quad \theta^\lambda_\lambda = 0\,. \tag{3.6}$$

Eqs. (3.4), (3.5) and (3.6) are necessary and sufficient to guarantee that the new tensor $\hat\theta_{\mu\nu}$ defined as

$$\hat\theta_{\mu\nu} = \theta_{\mu\nu} + \partial^\varrho \mathscr{G}_{[\varrho\mu]\nu} \tag{3.7}$$

is symmetric, by virtue of Eq. (3.4), and generates some charges, $\hat P_\mu$ and $\hat M_{\mu\nu}$ which, by Eq. (3.5), coincide with P_μ and $M_{\mu\nu}$. Note in addition that $\hat\theta^\lambda_\lambda = 0$ for $\theta^\lambda_\lambda = 0$ because of Eq. (3.6). In fact one has

$$\hat P_\mu - P_\mu = \int d^3x\, \partial^i \mathscr{G}_{[i,0]\mu} = 0$$

and

$$\begin{aligned}\hat M_{\mu\nu} - M_{\mu\nu} &= \int d^3x \{\partial^\varrho (x_\mu \mathscr{G}_{[\varrho,0]\nu} - x_\nu \mathscr{G}_{[\varrho,0]\mu}) - \mathscr{G}_{[\mu,0]\nu} + \mathscr{G}_{[\nu 0]\mu}\} \\ &= \int d^3x (\mathscr{G}_{[\nu,0]\mu} - \mathscr{G}_{[\mu,0]\nu})\end{aligned}$$

and such a difference vanishes only and only if Eq. (3.5) is valid. We note that only Eq. (3.5) must be assumed: indeed Eq. (3.4) follows if Eq. (3.5) holds.

For $\hat{D}-D$ one obtains

$$\hat{D}-D=\int d^3x\{\partial^\varrho(x^\lambda \mathcal{G}_{[\varrho,0]\lambda})-\mathcal{G}^\lambda_{[\lambda,0]}\}$$

which vanishes whenever

$$\mathcal{G}_{[\varrho,\sigma]}{}^\sigma=\partial^\sigma F_{[\sigma,\varrho]} \tag{3.8}$$

a condition which clearly is independent from Eqs. (3.4) and (3.5). Similarly

$$\begin{aligned}\hat{K}_\mu-K_\mu&=\int d^3x\{\partial^\varrho[(2x_\mu x^\lambda-g^\lambda_\mu x^2)\mathcal{G}_{[\varrho,0]\lambda}\\&\qquad-2(x^\lambda\mathcal{G}_{[\mu,0]\lambda}+x_\mu\mathcal{G}^\lambda_{[\lambda,0]}-x^\lambda\mathcal{G}_{[\lambda,0]\mu})\}\\&=2\int d^3x(-x_\mu\partial^\sigma F_{[\sigma,0]}(x)+x^\lambda\partial^\sigma F_{[\sigma,0][\mu,\lambda]}(x))\\&=2\int d^3x\{\partial^\sigma(x^\lambda F_{[\sigma,0][\mu,\lambda]}-x_\mu F_{[\sigma,0]})-F_{[\lambda,0][\mu,\lambda]}+F_{[\mu,0]}\}\\&=2\int d^3x\{-F_{[\lambda,0][\mu,\lambda]}+F_{[\mu,0]}\}\end{aligned}$$

where we have inserted Eqs. (3.8) and (3.5). To have $\hat{K}_\mu=K_\mu$ one has to require

$$F_{[\varrho,\sigma][\mu,\sigma]}=F_{[\varrho,\mu]}+\partial^\tau F_{[\tau,\varrho]\mu} \tag{3.9}$$

and vice versa. We note that Eq. (3.8) follows from Eqs. (3.9) and (3.5). The latter equations, together with the condition in Eq. (3.6), can be shown to admit non trivial solutions for $G_{[\varrho,\mu]\nu}$ in explicit examples, which shall not be described here.

3.2. Equal Time Commutators among the Conformal Charges

We want to compute the commutators among the charges P_μ, D, $M_{\mu\nu}$, K_μ. When the symmetry is exact the commutators are those of $SO(4,2)$. For broken symmetry the commutators among P_μ and $M_{\mu\nu}$ remain of course unchanged, while we expect the other commutators to be modified by terms depending on the breaking (we are excluding spontaneous breaking) [142, 143, 96–98, 166, 61, 48]. A first class of commutators to be evaluated is that of the commutators of D and K_μ with the Lorentz generators P_μ and $M_{\mu\nu}$. We can calculate them from Eqs. (3.2), since we know how P_μ and $M_{\mu\nu}$ act on $\theta_{\mu\nu}$ (that is $\theta_{\mu\nu}$ preserves its transformation properties under the Lorentz group). For instance,

$$\begin{aligned}[P_\mu,D]&=-i\int d^3x\,x^\nu[P_\mu,\theta_{0\nu}]=-i\int d^3x\,x^\nu(\partial\theta_{0\nu}/\partial x^\mu)\\&=-i\int d^3x\,\partial(x^\nu\theta_{0\nu})/\partial x^\mu+i\int d^3x\,g^\nu_\mu\theta_{0\nu}\\&=iP_\mu-ig_{\mu 0}\int d^3x\,\partial(x^\nu\theta_{0\nu})/\partial x^\mu=iP_\mu-g_{\mu 0}\int d^3x\,\theta^\lambda_\lambda\,.\end{aligned}$$

One can calculate quite similarly the commutators of D with $M_{\mu\nu}$, and of K_μ with D and $M_{\mu\nu}$. Collecting the results one has:

$$[P_\mu, D] = iP_\mu - ig_{\mu 0} \int d^3x\theta^\varrho_\varrho, \tag{3.10}$$

$$[M_{\mu\nu}, D] = -i \int d^3x(g_{\nu 0}x_\mu - g_{\mu 0}x_\nu)\,\theta^\varrho_\varrho, \tag{3.11}$$

$$[P_\nu, K_\mu] = 2ig_{\mu\nu}D + 2iM_{\mu\nu} - 2ig_{\nu 0} \int d^3x x_\mu \theta^\varrho_\varrho, \tag{3.12}$$

$$[M_{\varrho\sigma}, K_\mu] = i(g_{\sigma\mu}K_\varrho - g_{\varrho\mu}K_\sigma) - 2i \int d^3x(g_{\sigma 0}x_\varrho - g_{\varrho 0}x_\sigma)\,x_\mu \theta^\varrho_\varrho. \tag{3.13}$$

One sees that when $\theta^\varrho_\varrho \to 0$ one reobtains the commutators of the symmetric limit.

It is interesting to remark the loss of tensor character for D and K_μ (D is now no longer a scalar and K_μ is no longer a four-vector). At the same time, quite symmetrically, P_μ, for instance, does not behave any longer as a quantity of scale dimension $+1$ (in energy), as shown from Eq. (3.10). Clearly the loss of tensor character of D and K_μ is connected to the fact that they are no longer conserved generators. For instance Eq. (3.10) gives

$$[H, D] = iH - i \int d^3x\theta^\varrho_\varrho$$

and for $\theta^\varrho_\varrho \to 0$ one has $dD/dt = 0$

$$dD/dt = \partial D/\partial t + i[H, D] = \int d^3x\theta_{00} + i[H, D] = H + i[H, D] = 0\,.$$

We now calculate the commutators of D, and K_μ with themselves. For this calculation we cannot just commute the expressions in Eq. (3.2) with each other since we do not know how $\theta_{\mu\nu}$ behaves under D and K_μ when the symmetry is broken. The procedure to be followed here is taken from Ref. [41] and [76]. It is convenient to introduce a linear differential operator Δ_μ which acts on an operator $F(x)$ as follows

$$\Delta_\mu F(x) = dF(x)/dx_\mu + i[F(x), P_\mu]\,. \tag{3.14}$$

The operator Δ_μ has been so defined such as to differentiate only with respect to the explicit coordinate dependence. One easily derives

$$\Delta_\mu D = P_\mu\,, \tag{3.15}$$

$$\Delta_\mu K_\varrho = 2(M_{\varrho\mu} + g_{\varrho\mu}D)\,. \tag{3.16}$$

From Eqs. (3.15), (3.16), (3.11) and (3.12) one obtains

$$\Delta_\mu [D, K_\varrho] = 2i(M_{\gamma\mu} + g_{\varrho\mu}D - g_{\varrho 0} \int d^3x x_\mu \theta^\lambda_\lambda(x))$$

and, after integrating,

$$[D(x_0), K_\varrho(x_0)] = iK_\varrho(x_0) + iC_\varrho(x_0) - ig_{\varrho 0} \int d^3x x^2 \theta^\lambda_\lambda(x)\,. \tag{3.17}$$

In Eq. (3.17) C_ϱ is a constant of integration, satisfying

$$\Delta_\mu C_\varrho = 0 . \tag{3.18}$$

It has dimensions -1 in length (these are simply physical dimensions); similarly from (3.16), (3.13) and (3.17) one gets

$$\Delta_\mu [K_\varrho, K_\sigma] = 2i \int d^3x \{g_{\varrho 0}(2x_\mu x_\sigma + g_{\mu\sigma} x^2) + g_{\sigma 0}(2x_\mu x_\varrho + g_{\mu\varrho} x^2)\} \cdot \theta^\lambda_\lambda(x) + 2i(g_{\mu\varrho} C_\sigma - g_{\mu\sigma} C_\varrho)$$

which upon integration gives

$$[K_\varrho(x_0), K_\sigma(x_0)] = 2i \int d^3x\, x^2 (x_\sigma g_{\varrho 0} - x_\varrho g_{\sigma 0})\, \theta^\lambda_\lambda(x) + 2i N_{\varrho\sigma}(x) \tag{3.19}$$

where $N_{\varrho\sigma}$ satisfies

$$\Delta_\mu N_{\varrho\sigma} = g_{\mu\varrho} C_\sigma - g_{\sigma\mu} C_\varrho \tag{3.20}$$

and has physical dimensions $+2$.

We shall next take advantage of the Jacobi identities involving one Poincaré generator together with $K_\mu(x_0)$ and either $D(x_0)$ or $K_\mu(x_0)$. Such Jacobi identities are expected to hold even in the broken symmetry because at least one of the three charges are time independent and the equal time can be unambiguously defined. By deriving Eq. (3.17) with respect to x_0 (which is like using the Jacobi identity with P_μ) one has

$$\int d^3x \{[\theta(x), K_\mu(x_0)] - 2x_\mu [\theta(x), D(x_0)] = i dC_\mu(x_0)/dx_0 + 2i \int d^3x\, x_\mu \theta(x) - i g_{\mu 0} \int d^3x\, \partial_0(x^2 \theta(x))$$

and after using the transformation properties of $\theta(x)$ under $D(x_0)$ and $K_\mu(x_0)$ one obtains

$$\partial C_\mu(x_0)/\partial x_0 = \int d^3x \left\{ \sum_j (2\Delta_j x_\mu + 2x^\lambda \partial_\lambda - x^2 \partial_\lambda - 2\Delta_j x_\mu - 2x^\lambda \partial_\lambda) \cdot \delta_j u_j(x) - 2x_\mu \theta(x) + g_{\mu 0} \partial_0(x^2 \theta(x)) \right\} = 0$$

which gives

$$\partial C_\mu(x_0)/\partial x_0 = 0 . \tag{3.21}$$

From the Jacobi identity between $M_{\varrho\sigma}$, $D(x_0)$, and $K_\mu(x_0)$, after insertion of Eqs. (3.17), (3.11) and (3.13), and some algebraic steps one similarly obtains

$$[M_{\varrho\sigma}, C_\mu] = i(g_{\mu\sigma} C_\varrho - g_{\nu\varrho} C_\sigma) . \tag{3.22}$$

We shall now follow a standard procedure in similar cases and write, on the basis of the Eqs. (3.21) and (3.22),

$$C_\mu = \int d^3x\, t_{0\mu}(x) \tag{3.23}$$

where the local quantity $t_{\mu\nu}(x)$ satisfies

$$\partial^{\nu} t_{\nu\mu}(x)=0\,. \tag{3.24}$$

It does not explicitly depend on x_{μ} (that is, one has the equation $\Delta_{\varrho} t_{\mu\nu}=0$) and has physical dimensions -2.

One next takes the time derivative of Eq. (3.19) and makes use of the transformation properties of $\theta(x)$ with respect to $K_{\mu}(x_0)$, obtaining after some algebra the result

$$\mathrm{d}N_{\mu\nu}/\mathrm{d}x_0=0\,. \tag{3.25}$$

The next step is to take the Jacobi identity among $M_{\varrho\sigma}$, $K_{\mu}(x_0)$, and $K_{\nu}(x_0)$, and make use of Eqs. (3.19) and (3.11). One obtains in this way the result

$$[M_{\varrho\sigma}, N_{\mu\nu}]=i(g_{\varrho\nu}N_{\sigma\mu}-g_{\varrho\mu}N_{\sigma\nu}-g_{\sigma\nu}N_{\varrho\mu}+g_{\sigma\mu}N_{\varrho\nu})\,. \tag{3.26}$$

Eqs. (3.25) and (3.26) allows us to put $N_{\varrho\sigma}$ in the form

$$N_{\varrho\sigma}=\int \mathrm{d}^3x\, n_{0\varrho\sigma}(x) \tag{3.27}$$

where the local operator $n_{\tau\varrho\sigma}(x)$ is divergenceless in τ

$$\partial^{\tau} n_{\tau\varrho\sigma}(x)=0\,. \tag{3.28}$$

From Eqs. (3.20) and (3.23) one has moreover

$$n_{0\varrho\sigma}(x)=x_{\varrho}t_{0\sigma}-x_{\sigma}t_{0\varrho}+S_{0\varrho\sigma} \tag{3.29}$$

where $S_{0\varrho\sigma}$ does not contain any explicit dependence on x (i.e. it satisfies $\Delta_{\mu}S_{0\varrho\sigma}=0$). From Eqs. (3.28), (3.29) and (3.24) we obtain

$$\partial^{\tau} S_{\tau\varrho\sigma}=t_{\sigma\varrho}-t_{\varrho\sigma}\,. \tag{3.30}$$

We can introduce, following the well-known procedure by Møller [131] and Belinfante [9, 10] (note how the formal situation is here very similar although the physical content is different) a symmetric $t^{s}_{\varrho\sigma}(x)$ such that

$$C_{\varrho}=\int \mathrm{d}^3x\, t^{s}_{0\varrho}(x)\,, \tag{3.31}$$

$$N_{\varrho\sigma}=\int \mathrm{d}^3x\,[x_{\varrho}t^{s}_{0\sigma}(x)-x_{\sigma}t^{s}_{0\varrho}(x)]\,. \tag{3.32}$$

The new tensor $t^{s}_{\varrho\sigma}(x)$ is defined as (in strict analogy to the discussion of the preceeding chapter)

$$t^{s}_{\mu\nu}(x)=t_{\mu\nu}(x)+\partial^{\varrho}g_{[\varrho\mu]}(x)$$

$$g_{[\varrho\mu]\nu}(x)=\tfrac{1}{2}(S_{\varrho[\mu\nu]}(x)+S_{\mu[\nu\varrho]}(x)+S_{\nu[\mu\varrho]}(x))$$

and it satisfies the equations

$$\partial^{\varrho}t^{s}_{\varrho\sigma}(x)=0 \tag{3.33}$$

together with the symmetry condition

$$t^s_{\varrho\sigma}(x) = t^s_{\sigma\varrho}(x) .$$

The physical dimensions of $t^s_{\varrho\sigma}(x)$ are clearly -1. We also note that, when $\theta^\lambda_\lambda \to 0$, $t^s_{\varrho\sigma}(x)$ has to vanish.

We have thus completed the presentation of the formal consequences of the particular breaking scheme on the equal time commutation relations among conformal charges. We have in fact obtained the commutators in Eqs. (3.10)–(3.13), by an obvious reasoning, and by rather more sophisticated arguments we have derived the remaining commutators in Eqs. (3.17) and (3.19), with C_ϱ and $N_{\varrho\sigma}$ given through Eqs. (3.31) and (3.32) in terms of a tensor field $t^s_{\varrho\sigma}(x)$, which is symmetric, divergenceless, of physical dimensions -1, and vanishes when $\theta^\lambda_\lambda \to 0$.

We can further refine our argument by showing that such field cannot exist in any reasonable model. The argument runs as follows. Assume, for simplicity and temporarely, that the theory has only one independent mass parameter m such that $\theta^\lambda_\lambda \to 0$ is equivalent to $m \to 0$; and that for $m \to 0$ one has

$$m^{-\alpha} t^s_{\mu\nu} \to t^{(0)s}_{\mu\nu} \neq 0 \quad \text{and} \quad < \infty \tag{3.35}$$

where α is a real positive number. The operator $t^{(0)}_{\mu\nu}$ is symmetric, conserved, and is expected to transform covariantly under conformal transformations. It must than either be a c-number or have scale dimensions -4. Only in such case is indeed the equation $\partial^\mu t^{(0)s}_{\mu\nu} = 0$, Eq. (3.33), covariant under special conformal transformations. This is a very special case of a general theorem [63, 67] (the theorem of spin-dimension correlation). Note that the possibility that $t^{(0)s}_{\mu\nu}$ be a derivative of a covariant operator must be excluded since it would require for the latter scale dimensions > -1, in contrast to the positivity requirement on Lehman's spectral function. From Eq. (3.35) we see that the dimensions of $t^{(0)s}_{\mu\nu}$ are $-2+\alpha > 2$. The only escape is then that $t^{(0)s}_{\mu\nu}$ be a c-number: but then to have $[D(x_0), K_\mu(x_0)]$ well-defined it would have to vanish (one assumes $C_\varrho < \infty$). In conclusion Eq. (3.35) cannot be satisfied and $t^{(s)}_{\mu\nu}(x)$ must vanish.

Finally, we have to remove the condition of a single mass parameter. If there are N independent mass parameters m_i one assumes that for $m_i \to 0$ a finite limit exists for some product $f(m_1, \ldots, m_N)^{-1} t^{(s)}_{\mu\nu}(x)$, where $f(m_1, \ldots, m_N)$ is some function of the N-masses. The argument then goes as before. So one can conclude, rather generally, that the commutators are those in Eqs. (3.10)–(3.13) together with

$$[D(x_0), K_\mu(x_0)] = iK_\mu(x_0) - ig_{\mu 0} \int d^3x\, x^2\, \theta(x) , \tag{3.38}$$

$$[K_\mu(x_0), K_\nu(x_0)] = 2i \int d^3x\, x^2 (x_\nu g_{\mu 0} - x_\mu g_{\nu 0})\, \theta(x) . \tag{3.39}$$

In terms of the commutators $[\theta_{\mu\nu}, \theta_{\varrho\sigma}]$ the validity of these Eqs. (3.10) to (3.13) and (3.38), (3.39) is equivalent to a set of conditions on the Schwinger terms. Included in such a set of conditions is the well-known condition by Schwinger for relativistic covariance [146]. We shall not here further discuss such a subject. Additional discussion can be found in Ref. [40]. We want here to mention another interesting consequence of the commutators (3.10)–(3.13) and (3.38), (3.39). It is easily seen that the charges $\boldsymbol{P}$, $\varepsilon_{ijk} M^{jk}$, D, and $\boldsymbol{K}$ satisfy the unmodified commutation relations even for broken symmetry. In fact they generate a non-symmetry group, which may be useful for instance in classification of states. The Hamiltonian, of course, does not commute with the full set of generators.

4. Restrictions from Conformal Covariance on Equal Time Commutators

4.1. Outline of the Problem

In this chapter we shall be interested in deriving the restrictions that conformal covariance implies on equal-time commutators of local fields. We assume that the equal-time commutator of two local fields $A(x)$ and $B(y)$ is a temperate distribution with support at $x=y$. This in fact follows from locality and the general hypothesis that the commutator be a temperate distribution on any space-like surface and that the equal time limit exists. It will be convenient to introduce a timelike four-vector n_μ of fixed components in any frame and to write the equal-time-commutator (briefly: etc.) in the form

$$\begin{aligned} &\delta[n\cdot(x-y)]\,[A(x), B(y)] = \sum_{k=0}^{N} i^k S^{\tau_1\ldots\tau_k}(A, B, n, y)\, \partial_{\tau_1}\ldots\partial_{\tau_k}\delta(x-y) \\ &\partial_{\tau_i} = \partial/\partial x^{\tau_i} \qquad \delta(x) = \delta^{(4)}(x) \end{aligned} \tag{4.1}$$

where $S^{\tau_1\ldots\tau_k}(A, B, n, y)$ are linear combinations of local operators and depend on A and B. The coefficients in such linear combinations are homogeneous of degree -1 in n_μ, as evident from (4.1). Also $S^{\tau_1\ldots\tau_k}$ is completely symmetric and the relations $n_{\tau_i} S^{\tau_1\ldots\tau_k}=0$ hold for each i, such that ∂_{τ_i} only acts on a component transverse to n_μ. We shall sometime abbreviate $S^{\tau_1\ldots\tau_k}$ into $S^{(k)}$.

The conformal charges will be taken in the form

$$P_\lambda = \int d^4x\,\delta(nx)\, n^\mu \theta_{\mu\lambda} \qquad M_{\mu\nu} = \int d^4x\,\delta(nx)\, n^\varrho (x_\mu \theta_{\varrho\nu} - x_\nu \theta_{\varrho\mu}), \tag{4.2}$$

$$D = \int d^4x\,\delta(nx)\, n^\varrho x^\sigma \theta_{\varrho\sigma} \qquad K_\mu = \int d^4x\,\delta(nx)\, n^\varrho (2x_\mu x^\sigma \theta_{\varrho\sigma} - x^2 \theta_{\varrho\mu}). \tag{4.3}$$

A local field $A_r(x)$, where r denotes a set of spinor or tensor indices, that transforms covariantly under the algebra satisfies the relations (see § 2) (we assume $k_\lambda = 0$ for such a field)

$$[P_\lambda, A_r(x)] = -i\partial_\lambda A_r(x)\,, \tag{4.4}$$

$$[M_{\mu\nu}, A_r(x)] = -i[(x_\mu\partial_\nu - x_\nu\partial_\mu)\,\delta_r^s - i(\Sigma_{\mu\nu})_r^s]\,A_s(x)\,, \tag{4.5}$$

$$[D, A_r(x)] = i(l_A - x^\varrho\partial_\varrho)\,A_r(x)\,, \tag{4.6}$$

$$[K_\mu, A_r(x)] = i[(2l_A x_\mu - 2x_\mu x^\varrho\partial_\varrho + x^2\partial_\mu)\,\delta_r^s + 2ix^\varrho(\Sigma_{\mu\varrho})_r^s]\,A_s(x)\,. \tag{4.7}$$

Of course P_λ and $M_{\mu\nu}$, being the generators of the Lorentz group, are conserved; and Eqs. (4.4) and (4.5) are just the expressions of Lorentz covariance of the field $A_r(x)$. Thus some of our calculations, in so far as Eqs. (4.3), (4.6) and (4.7) are not used have only to do with the usual Lorentz covariance, and not at all with the conformal group in itself. However it would be rather difficult to treat the consequences of covariance under D and K_μ, without having first summarized, at least, the consequences of covariance under P_μ and $M_{\mu\nu}$. The restrictions which follow from Eqs. (4.4)–(4.7) are, strictly, a number of relations to be satisfied by the Schwinger terms in the commutator $[\theta_{\mu\nu}, A_r]$; in our notations of Eq. (4.1), they are restrictions on $S^{\tau_1\ldots\tau_k}(\theta_{\mu\nu}, A_r, n, x)$. Additional restrictions follow however from use of the Jacobi identities among two local operators $A_r(x)$ and $B_r(x)$ and a generator P_μ, $M_{\mu\nu}$, D, or K_μ. Such restrictions directly limit the forms for $S^{\tau_1\ldots\tau_k}(A_r, B_s, n, x)$. However one must stress that, whereas such additional restrictions are rigorous when the Jacobi identity involves P_μ or $M_{\mu\nu}$, which are time independent (and therefore can be easily translated to the limiting value $x_0 = y_0$), when dealing with D and K_μ the consequences hold only for exact symmetry. As we have said, before dealing with the derivation of the implications of conformal symmetry we must summarize some general properties of etc. and in particular discuss the more general consequences of Lorentz invariance alone. We shall follow here the treatment in Ref. [40], where a consistent notation and derivation is developed.

4.2. Summary of General Properties of Equal Time Commutators

We shall first deal with the so-called property of integrability for Schwinger terms. It will be convenient to introduce the time ordered product along n_μ

$$T_n(A(x)\,B(0)) = \theta(nx)\,A(x)\,B(0) + \theta(-nx)\,B(0)\,A(x) \tag{4.8}$$

from which

$$(\mathrm{d}/\mathrm{d}n_\mu)\,T_n(A(x)\,B(0)) = x^\mu\,[A(x), B(0)]\,\delta(nx)\,. \tag{4.9}$$

Therefore for $k \leq N$ one has

$$(\mathrm{d}/\mathrm{d}n_\mu) \int \mathrm{d}^4x\, x^{\mu_1} \ldots x^{\mu_n} T_n(A(x)\, B(0)) \\ = \sum_{l=0}^{N} i S^{\tau_1 \ldots \tau_l}(A, B, n, 0) \int \mathrm{d}^4x\, x^{\mu_1} \ldots x^{\mu_k} \partial_{\tau_1} \ldots \partial_{\tau_l} \delta(x) \tag{4.10}$$

giving

$$S^{\mu_1 \ldots \mu_{k+1}}(A, B, n, 0) \\ = -i(\mathrm{d}/\mathrm{d}n_{\mu_{k+1}}) \int \mathrm{d}^4x (ix^{\mu_1}) \ldots (ix^{\mu_k})\, T_n(A(x)\, B(0)) . \tag{4.11}$$

One next makes use of the symmetry of $S^{\mu_1 \ldots \mu_{k+1}}$ and obtains

$$S^{\mu_1 \ldots \mu_{k+1}}(A, B, n, x) = (\mathrm{d}/\mathrm{d}n_{\mu_1} \ldots \mathrm{d}/\mathrm{d}n_{\mu_k})\, Z^{(k)}(A, B, n, x) \tag{4.12}$$

where the symbol $Z^{(k)}$ is defined apart from an additive polynomial in n_μ of degree $k-1$. If will be convenient to choose such polynomial as homogeneous of degree $k-1$ in n_μ, so that $Z^{(k)}$ is homogeneous of degree $k-1$. Note that covariance of $T_n(A(x)\, B(0))$ is equivalent to absence of all $S^{(k)}$.

Under the exchange $A \leftrightarrow B$ one obviously has:

$$[A(x), B(0)] = -[B(0), A(x)] .$$

This property implies the relation

$$S^{\tau_1 \ldots \tau_j}(A, B, n, x) + (-i)^j S^{\tau_1 \ldots \tau_j}(B, A, n, x) \\ = -\sum_{k=j+1}^{N} i^{k+j} \binom{k}{j} \partial_{\tau_{j+1}} \ldots \partial_{\tau_k} S^{\tau_1 \ldots \tau_k}(B, A, n, x) . \tag{4.13}$$

Eq. (4.13) fixes in $S^{(j)}$ the term of parity $(-1)^j$ under $A \leftrightarrow B$ in terms of $S^{(j+1)}$, $S^{(j+2)}$, etc. In particular (for $j = N$)

$$S^{\tau_1 \ldots \tau_N}(A, B, n, x) = (-1)^{N+1} S^{\tau_1 \ldots \tau_N}(B, A, n, x) . \tag{4.14}$$

For the derivation of Eq. (4.3) (see Ref. [41]).

An additional, quite obvious, relation follows directly from Eq. (4.1). We introduce the abbreviation for any vector V_μ

$$V_{\bar{\mu}} = V_\mu - (nv/n^2)\, n_\mu \tag{4.15}$$

and call $\bar{g}_{\mu\nu}$ the symbol

$$\bar{g}_{\mu\nu} = g_{\mu\nu} - (n_\mu n_\nu / n^2) = g_{\bar{\mu}\nu} = g_{\mu\bar{\nu}} = g_{\bar{\mu}\bar{\nu}} . \tag{4.16}$$

One then has for the "transverse" derivatives

$$S^{(0)}(\partial_{\bar{\varrho}} A, B, n, x) = 0 , \tag{4.17}$$

$$S^{\tau_1 \ldots \tau_k}(\partial_{\bar{\varrho}} A, B, n, x) = k^{-1} \sum_{i=1}^{k} \bar{g}_\varrho^{\tau_i} S^{\tau_1 \ldots \hat{\tau}_i \ldots \tau_k}(A, B, n, x) \tag{4.18}$$

(the notation $\tau_1 \ldots \hat{\tau}_i \ldots \tau_k$ means that the index τ_i is omitted from the sequence $\tau_1 \ldots \tau_k$).

Finally we have a very important theorem which follows from *PCT*. We shall limit ourselves here in enunciating the theorem, without reporting the proof:

PCT Theorem on Equal-Time-Commutators. In the expansion in Eq. (4.1) for an equal time commutator, the only singularities in n_μ of each $S^{\tau_1 \ldots \tau_k}$ are poles at $n^2 = 0$.

The theorem results in powerful restrictions on the nature of the Schwinger terms. It is particularly useful when applied in conjunction with the other general properties of etc., in particular with the condition of homogeneity. As a consequence, we mention the following "even-odd rule for Schwinger terms". Let us call $S^{\tau_1 \ldots \tau_k}_{\mu_1 \ldots \mu_j, \nu_1 \ldots \nu_l}$ (briefly $S^k_{j,l}$) the k-th Schwinger term for the commutator $[A_{\mu_1 \ldots \mu_j}(x, 0), B_{\nu_1 \ldots \nu_l}(0)]$; $S^k_{j,l}$ is a sum of local operators. Such operators all have the total number of tensor indices even or odd. The case "even" occurs for $j+k+l=$ odd, the case "odd" for $j+k+l=$ even. Example: An equal time commutators of two scalar object can only give vectors, third rank tensors, etc., plus Schwinger terms.

4.3. Restrictions from Lorentz Covariance on Equal-Time Commutators

From Eqs. (4.2), (4.4), (4.5) one derives formally [41]

$$n^\varrho S^{(0)}(\theta_{\varrho\sigma}, A_r, n, x) = -i\partial_\sigma A_r(x) \tag{4.19}$$

and

$$n^\varrho [S^\mu(\theta_\varrho, A_r, n, x) - S^\nu(\theta^\mu_\varrho, A_r, n, x)] = -i(\Sigma^{\mu\nu})^s_r A_s(x)\,. \tag{4.20}$$

Eqs. (4.19) and (4.20) are two independent limitations on the Schwinger terms and they express the content of Eqs. (4.4) and (4.5) for any energy-momentum tensor which is symmetric, divergenceless, and related through Eqs. (4.2) to the generators P_μ and $M_{\mu\nu}$ (such as for instance the Belinfante-Møller $\theta_{\mu\nu}$). One finds from Eqs. (4.19) and (4.20) that the commutator of θ_{00} with any local operator has the form

$$\begin{aligned}\delta(x_0 - y_0)\,[\theta_{00}(x), A_r(y)] \\ = -i[(\partial/\partial y_0) A_r(y)]\,\delta(x-y) + (\Sigma^{\mu 0})^s_r A_s(y)(\partial/\partial y_\mu)\delta(x-y) + \cdots.\end{aligned} \tag{4.21}$$

For the interesting case of $[\theta_{00}, \theta_{00}]$ one obtains in particular the well-known theorem of Schwinger [146]. One also reobtains the results of Boulware and Deser [17] about the etc. $[\theta_{\mu\nu}(x), \theta_{\varrho\sigma}(0)]$. Here one has to make use of Eq. (4.13) and of the local conservation of $\theta_{\mu\nu}$. Additional restrictions are implied by the Jacobi identities for $A_r(x)$, $B_s(x)$, and either P_μ or $M_{\mu\nu}$. One obtains

$$\partial_\mu S^{\tau_1\ldots\tau_k}(A_r, B_s, n, x) = S^{\tau_1\ldots\tau_k}(\partial_\mu, A_r, B_s, n, x) + S^{\tau_1\ldots\tau_k}(A_r, \partial_\mu, B_s, n, x) \tag{4.22}$$

from the Jacobi identity with P_μ, and by a slightly more complicate algebra [41], the relation:

$$\begin{aligned}(k+1)\,[S^{\mu,\tau_1\ldots\tau_k}(\partial^\nu A_r, B_s, n, x) - S^{\nu,\tau_1\ldots\tau_k}(\partial^\mu A_r, B_s, n, x)] \\ = [S^{\tau_1\ldots\tau_k}(A_r, B_s, n, x), M^{\mu\nu}] - S^{\tau_1\ldots\tau_k}((\Sigma^{\mu\nu} A)_r, B_s, n, x) \\ - S^{\tau_1\ldots\tau_k}(A_r, (\Sigma^{\mu\nu} B)_s, n, x)\,.\end{aligned} \tag{4.23}$$

One sees from Eq. (4.23) that the Schwinger terms of order $k+1$ in a commutator involving a derivative are related to those of order k without derivative. When Eq. (4.23) is taken together with Eqs. (4.17) and (4.18) one obtains a complete determination of $S^{\tau_1\ldots\tau_{k+1}}(\partial_\mu A, B, n, x)$ in terms of $S^{\tau_1\ldots\tau_k}(A, B, n, x)$. In fact it can be shown [144] that one can write the following compact expression, equivalent to the set of Eqs. (4.17), (4.18) and (4.13)

$$\begin{aligned} &iS^{\tau_1\ldots\tau_k}(\partial^\nu A, B, n, x) \\ &= k^{-2} \sum_{j=1}^{k} [(k-1)\, g^{\nu\tau_j} + (\mathrm{d}n^\nu/\mathrm{d}n_{\tau_j})]\, S^{\tau_1\ldots\hat{\tau}_j\ldots\tau_k}(A, B, n, x)\,.\end{aligned} \tag{4.24}$$

The result contained in Eq. (4.24) was first derived by a different method [155, 42]. The implications of Lorentz covariance on etc. have also been discussed by Brown, Gross and Jackiw [27, 28, 99, 85, 86]. Eq. (4.24) can be given a rather simple form in terms of $Z^{(k)}$, see Eq. (4.12). One finds

$$ikZ^{(k)}(\partial^\nu A, B, n, x) = n\,Z^{(k-1)}(A, B, n, x) + P^\nu(n) \tag{4.25}$$

where $P^\nu(n)$ is a homogeneous, but otherwise unspecified, polynomial of degree $k-1$ in n.

4.4. Restrictions on Equal Time Commutators Ensuing from Conformal Symmetry

The equations

$$[D, A_r(x)] = i(l_A - x^\varrho\,\partial_\varrho)\, A_r(x)\,, \tag{4.26}$$

$$[K_\mu, A_r(x)] = i\,[(2l_A x_\mu - 2x_\mu x^\varrho\,\partial_\varrho)\,\delta_r^s + 2ix^\varrho(\Sigma_{\mu\varrho})_r^s]\, A_s(x)\,, \tag{4.27}$$

express the assumption that $A_r(x)$ transforms according to an irreducible representation of the conformal algebra with $k_\mu = 0$, finite dimensional with respect to the little group, as reviewed in Chapter 2. We recall that

$$D = \int d^4x\delta(nx)\, n^\varrho x^\sigma \theta_{\varrho\sigma}, \tag{4.28}$$

$$K_\mu = \int d^4x\delta(nx)\, n^\varrho(2x_\mu x^\sigma \theta_{\varrho\sigma} - x^2\theta_{\varrho\mu}). \tag{4.29}$$

One immediately obtains:

$$\begin{aligned}[D, A_r(0)] &= \int d^4x\delta(nx)\, n^\varrho x^\sigma [\theta_{\varrho\sigma}(x), A_r(0)] \\ &= \sum_k i^k n^\varrho S^{\tau_1\ldots\tau_k}(\theta_{\varrho\sigma}, A_r, 0) \int d^4x x^\sigma \partial_{\tau_1}\ldots\partial_{\tau_k}\delta(x) \\ &= -in^\varrho S^\sigma(\theta_{\varrho\sigma}, A_r, 0)\end{aligned}$$

and after using Eq. (4.26) one can write:

$$n^\varrho S^\sigma(\theta_{\mu\sigma}, A_r, n, x) = -l_A A_r(x). \tag{4.30}$$

With K_μ one proceeds by similar steps. One has

$$\begin{aligned}[K_\mu, A_r(0)] &= \sum_k i^k n^\varrho \int d^4x\, \{S^{\tau_1\ldots\tau_k}(\theta_{\varrho\sigma}, A_r, 0)\, 2x_\mu x^\sigma \\ &\quad - S^{\tau_1\ldots\tau_k}(\theta_{\varrho\mu}, A_r, 0)\, x^2\}\, \partial_{\tau_1}\ldots\partial_{\tau_k}\delta(x) \\ &= -2n^\varrho\{2g_{\mu\tau_1}, g^\sigma_{\tau_2} S^{\tau_1\tau_2}(\theta_{\varrho\sigma}, A_r, 0) - g_{\tau_1\tau_2} S^{\tau_1\tau_2}(\theta_{\varrho\mu}, A_r, 0)\}.\end{aligned}$$

Then one compares with Eq. (4.27), obtaining

$$g_{\tau_1\tau_2} n^\varrho S^{\tau_1\tau_2}(\theta_{\varrho\mu}, A_r, n, x) = 2g_{\mu\tau_1} n^\varrho S^{\tau_1\tau_2}(\theta_{\varrho\tau_2}, A_r, n, x). \tag{4.31}$$

The main results are thus Eqs. (4.30) and (4.31). They express limitations on the Schwinger terms of $[\theta_{\mu\nu}, A_r]$.

The restrictions obtainable from the Jacobi identities are derived in the Appendix A. They are valid only if the symmetry is exact, as we have already explained. In fact D and K_μ are not in general independent of time and the Jacobi identity among A_r, B_s and D or K_μ cannot be justified a priori.

5. Manifestly Conformal Covariant Structure of Space-Time

5.1. Spinor Operators

As well known [109, 122] the conformal algebra on space-time is isomorphic to the orthogonal $O(4,2)$ algebra. We recall that the 15 generators of $O(4,2)$, which are arranged into a skew-symmetric tensor J_{AB}, are given by

$$J_\mu = M_{\mu\nu}, \quad J_{5\mu} = \tfrac{1}{2}(P_\mu - K_\mu), \quad J_{6\mu} = \tfrac{1}{2}(P_\mu + K_\mu), \quad J_{65} = D \tag{5.1}$$

in terms of the generators of the conformal algebra.

In fact, from the commutation rules of the conformal algebra it follows

$$[J_{AB}, J_{CD}] = i(g_{AD}J_{BC} + g_{BC}J_{AD} - g_{AC}J_{BD} - g_{BD}J_{AC}) \tag{5.2}$$

and

$$g_{AA} = (+\,-\,-\,-,\,-\,+)\ A = 0, 1, 2, 3, 5, 6$$

$O(4,2)$ is a rank three algebra, with Casimir Operators

$$\begin{aligned} C_{\mathrm{I}} &= J^{AB}J_{AB}, \\ C_{\mathrm{II}} &= \varepsilon_{ABCDEF}J^{AB}J^{CD}J^{EF}, \\ C_{\mathrm{III}} &= J^B_A J^C_B J^D_C J^A_D. \end{aligned} \tag{5.3}$$

The irreducible representations of the algebra are specified by the eigenvalues of these operators.

It is a simple task to prove that the action of the conformal group on the Minkowski space is equivalent to the action of $O(4,2)$ on the homogeneous space $O(4,2)/[O(3,1)\otimes D]$. Let us consider an arbitrary point P in the six-dimensional pseudo-orthogonal space $O(4,2)$, $P \equiv (\eta_A = g^B_A \eta_B)$; and choose as independent variables

$$x_\mu = k^{-1}\eta_\mu, \quad k = \eta_5 + \eta_6 \quad \text{and} \quad \eta^2 \quad (\eta^2 = \eta^A\eta_A). \tag{5.4}$$

A transformation $\Lambda \in O(4,2)$, acting as $\eta^A = \Lambda^A_B \eta^B$, induces on the new variables the following transformations [143]

$$\left.\begin{aligned} x'_\mu &= L^\nu_\mu x_\nu + a_\mu \\ k' &= k \end{aligned}\right\} \quad \text{(Poincaré transformations)}, \tag{5.5}$$

$$\left.\begin{aligned} x'_\mu &= e^\lambda x_\mu \\ k' &= e^{-\lambda} k \end{aligned}\right\} \quad \text{(dilatations)}, \tag{5.6}$$

$$\begin{aligned} x'_\mu &= \frac{x_\mu + c_\mu(x^2 - \eta^2/k^2)}{1 + 2c\cdot x + c^2(x^2 - \eta^2/k^2)} \\ &\qquad \text{(special conformal transformations)} \\ k' &= k(1 + 2c\cdot x + c^2(x^2 - \eta^2/k^2)). \end{aligned} \tag{5.7}$$

We then see that, in order to recover the conformal transformations laws in space-time, the point $x \in M_4$ (M_4 = Minkowski space) has to be identified with the set $x_\mu \equiv \{\varrho\eta_\mu, \varrho k\}$ ϱ arbitrary, on the cone $\eta^2 = 0$.

Let us now consider operator-valued spinor functions $\Psi_{\{\alpha\}}(\eta)$ defined on the cone $\eta^2 = 0$. They transform according to

$$\begin{aligned} \delta\Psi_{\{\alpha\}}(\eta) &= -i\varepsilon^{AB} J^{\{\beta\}}_{AB\{\alpha\}} \Psi_{\{\beta\}}(\eta) \\ &= -i\varepsilon^{AB}(L_{AB}\delta^{\{\beta\}}_{\{\alpha\}} + S^{\{\beta\}}_{\{\alpha\}})\,\Psi_{\{\beta\}}(\eta) \quad (L_{AB} = i(\eta_B\partial_A - \eta_A\partial_B)) \end{aligned} \tag{5.8}$$

where $S_{\{\alpha\}}^{\{\beta\}}$ is the matrix of an irreducible representation of the spinor group $SU(2,2)$, locally isomorphic to $O(4,2)$. We assume that these functions are homogeneous of degree λ [1], i.e.

$$\eta^A \partial_A \Psi_{\{\alpha\}}(\eta) = \lambda \Psi_{\{\alpha\}}(\eta) \qquad (\eta^A \partial_A = k\partial/\partial k). \tag{5.9}$$

The function

$$\Psi'_{\{\alpha\}}(x) = k^{-\lambda} \Psi_{\{\alpha\}}(\eta) \tag{5.10}$$

is then defined on space-time. However [122] the operator $\Psi'_{\{\alpha\}}(x)$ is non-local in the sense that

$$[\Psi'_{\{\alpha\}}(x), P_\mu] = (i\partial/\partial x_\mu + \pi_\mu)\, \Psi'_{\{\alpha\}}(x) \tag{5.11}$$

where $\pi_\mu = S_{6\mu} + S_{5\mu}$. From (5.11) one easily sees that the new operator-valued function

$$O_{\{\alpha\}}(x) = (e^{-ix\cdot\pi}\Psi')_{\{\alpha\}}(x) = k^{-\lambda}(e^{-ix\cdot\pi})_{\{\alpha\}}^{\{\beta\}}\, \Psi_{\{\beta\}}(\eta) \tag{5.12}$$

transforms according to a representation of the conformal algebra on space-time induced from a representation of the stability algebra at $x=0$ $\Sigma_{\mu\nu}, \Delta, K_\mu$ with matrices

$$\Sigma_{\mu\nu} = S_{\mu\nu}, \quad \Delta = S_{65} - i\lambda I, \quad K_\mu = S_{6\mu} - S_{5\mu}. \tag{5.13}$$

5.2. Irreducible Representations in Space-Time

In this section we are interested in investigating the structure of a particular family of irreducible representations of conformal algebra on space-time [132, 94, 62, 84, 77, 111, 139, 36, 140, 156]. We classify those representations which contain infinite towers of irreducible representations of $SL(2, C)$ of the type $[n/2, n/2]$; i.e. tensor representations.
The study of this particular class of representation is expecially motivated by the fact that they are relevant in operator product expansions, a problem that will be deeply investigated in the next sections.

5.3. Structure of Representations and Connections with $O(4,2)$ Covariant Tensors

The classification of the irreducible representations is achieved by making use of the following two basic Lemmas [67]:

[1] Note that in general, for $\eta^2 \neq 0$, $\eta^A \partial_A = k\,\partial/\partial k + 2\eta^2\, \partial/\partial\eta^2$ so Eq. (5.10) holds only for $\eta^2 = 0$. This is due to the fact that only the hypersurface $\eta^2 = 0$ is invariant under dilatations on the six-dimensional space:

$$\eta_A \to \lambda \eta_A .$$

Lemma 1. Every irreducible (infinite-dimensional) representation of conformal algebra which contains a ladder of Lorentz tensors of order $n+k$, $k=0,1,2\ldots$ (i.e. irreducible $SL(2,C)$ representations of the type $\frac{1}{2}(n+k)$, $\frac{1}{2}(n+k)$) is uniquely specified by an irreducible Lorentz tensor $(n/2, n/2)$ of given dimension l_n, annihilated by K_λ, i.e. by an irreducible representation of $SL(2,C)\otimes D$. The last assertion follows from the structure of the stability algebra.

The proof of Lemma 1 follows from the fact that the Casimir Operators (5.3) are given in such representations by

$$\begin{aligned} C_{\mathrm{I}} &= 2l_n(l_n-4)+2n(n+2) \\ C_{\mathrm{II}} &= 0 \\ C_{\mathrm{III}} &= n(n+2)\,[3+l_n(l_n-4)] \end{aligned} \tag{5.14}$$

as can be obtained by evaluating their eigenvalues on the lowest order tensor $(n/2, n/2)$ annihilated by K_λ[2]. Therefore these representations are specified by two numbers n, l_n where n is a non-negative integer and l_n assumes any value (with the exception $l_n=2+n$).

Lemma 2. Every irreducible representation of the conformal algebra, which according to the previous Lemma is uniquely specified by a Lorentz tensor of order n (annihilated by K_λ) and its dimension l_n, can be uniquely enlarged ($l_n \neq 2+n$), to a tensor representation of $O(4,2)$ acting on $O(4,2)/[IO(3,1)\otimes D]$.

The tensors $\Psi_{A_1\ldots A_n}(\eta)$ are specified [66, 67] by the following properties

They are homogeneous of degree $-l_n$. (5.15a)

They are irreducible, i.e. symmetric and traceless. (5.15b)

They satisfy two sets of supplementary conditions
$\eta^{A_1}\Psi_{A_1\ldots A_n}(\eta)=0$ and $\partial^{A_1}\Psi_{A_1\ldots A_n}(\eta)=0$ (5.15c)
(generalized Lorentz condition).

It is possible to prove (see Appendix 3) that these properties for the tensor $\Psi_{A_1\ldots A_n}$ are equivalent to the following ones.

The tensors are irreducible with respect to the orbital part of the algebra $O(4,2)$. (5.16a)

They are irreducible with respect to the spin part of the algebra $O(4,2)$. (5.16b)

They are irreducible with respect to the whole algebra, i.e. $L\cdot S$ is a constant on these representations. (5.16c)

---

[2] For example, the Casimir C_{I}, in terms of conformal generators turns out to be $C_{\mathrm{I}}=M_{\mu\nu}M^{\mu\nu}+2P\cdot K-2D^2+8iD$.

In Appendix C we prove Lemma 2 and exhibit the mapping between the tensors $\Psi_{A_1 \ldots A_n}$ and the irreducible representations of the type considered in Lemma 1. We find in particular that in $O_{A_1 \ldots A_n}$ indices 5 and 6 are interchangeable and that (α's and μ's are 0, 1, 2, 3)

$$O_{\alpha_1 \ldots \alpha_{n-k}, 6, 6, \ldots 6}(x) = 2^{-k} \frac{\Gamma(l_n - 2 - n)}{\Gamma(l_n - 2 - n + k)} \partial_{\mu_1} \ldots \partial_{\mu_k} O^{\mu_1 \ldots \mu_k}_{\alpha_1 \ldots \alpha_{n-k}}(x) . \tag{5.17}$$

5.4. "Canonical Dimensions"

We note that the above procedure is meaningless for $l_n = 2 + n$, i.e. in the case of "canonical dimensions". In this case, in fact, the second supplementary condition (5.15c) cannot be imposed unless the tensor $O_{\alpha_1 \ldots \alpha_n}(x)$ is conserved; but, in this case, it becomes an identity: it simply tells us that the conservation law $\partial^\alpha O_{\alpha \alpha_1 \ldots \alpha_{n-1}}(x) = 0$ is a conformal invariant equation. However the components $O_{\alpha_1 \ldots \alpha_J}$ are still indetermined and the previously one-to-one correspondence is no longer possible.

In the case $l_n = 2 + n$, but for nonconserved 4-tensors $O_{\alpha_1 \ldots \alpha_n}(x)$ the second supplementary condition is no longer valid. Moreover as a consequence of a theorem proved in Ref. [63] both tensors $O_{\alpha_1 \ldots \alpha_n}(x)$, $\partial^\alpha O_{\alpha \alpha_1 \ldots \alpha_{n-1}}(x)$ are annihilated by K_λ at $x = 0$, i.e. there is a degeneragy of the eigenspace corresponding to $K_\lambda = 0$. This pathology has a well defined counterpart with manifestly covariant tensors. In fact, given the non-conserved tensor $\Psi_{A_1 \ldots A_n}(\eta)$, one can always define:

$$\tilde{\Psi}_{A_1 \ldots A_n}(\eta) = \Psi_{A_1 \ldots A_n}(\eta) - \frac{1}{2} \sum_{i=1}^{n} \eta_{A_i} \partial^A \Psi_{A A_1 \ldots \hat{A}_i \ldots A_n}(\eta) \tag{5.18}$$

which is conserved, since the tensor $\partial^A \Psi_{A A_1 \ldots A_i \ldots A_n}(\eta)$ is conserved as it does not carry "canonical dimensions". This ensures that the tensor defined in (5.18) is irreducible under $O(4, 2)$ transformations [according to Eq. (5.15a)–(5.16c)]. We still observe that the tensor $\partial^A \Psi_{A A_1 \ldots A_{n-1}}(\eta)$ is a genuine 6-tensor, in the sense that it satisfies both supplementary conditions, so that its components can be evaluated with the method exploited above.

The conformally covariant Maxwell equation [109] is a simple example of the method sketched above. It reads as

$$\Box_6 A_c(\eta) = J_c(\eta) \tag{5.19}$$

(note that $\Box_6 = k^{-2} \Box_4 + 4(1 - l_A) \partial / \partial \eta^2$ is defined on the cone only for $l_A = 1$). Using Eq. (5.12) we get in space-time

$$\Box_4 A_\nu(x) + (2 i \pi \cdot \partial - \pi^2)^B_\nu A_B(x) = \Box_4 A_\nu(x) - \partial_\nu \partial^\mu A_\mu(x) = J_\nu(x) \tag{5.20}$$

where we used (5.17) to derive $A_5(x) = -\frac{1}{4} \partial^\mu A_\mu(x)$.

This is the Maxwell equation in a general gauge. In particular the Lorentz condition $\partial^\nu A_\nu(x)=0$ cannot be imposed as an operator equation, as it is not conformal invariant. Finally Eq. (5.19) defines $J_5(x)$ as in

$$J_5(x)=-\tfrac{1}{4}\Box_4\partial^\nu A_\nu(x) \tag{5.21}$$

so $J_c(x)$ is a genuine six-vector in accordance to the wave equation.

6. Conformal Invariant Vacuum Expectation Values

It is important to investigate the consequences of conformal symmetry on vacuum expectation values (V.E.V.) of operator products (n-point functions).

It is in this connection that the use of the manifestly covariant formalism appears as extremely powerful. It will be convenient however to also exploit the corresponding space-time formulation.

6.1. Two-Point Functions

Let us start with the simplest case of the two-point function. Consider the V.E.V. of two (conformal) scalar fields of dimension l_A, l_B respectively

$$\langle 0|A(\eta)\,B(\eta')|0\rangle=F(\eta\cdot\eta') \tag{6.1}$$

(Eq. (6.1) obviously follows from invariance under $O(4,2)$ transformations). With the previously derived parametrization of the six-dimensional cone $\eta^2=0$ (where $\eta_A=(x_\mu,k)$) the homogeneity conditions

$$\begin{aligned}\eta^A\partial_A F(\eta\cdot\eta')&=-l_A F(\eta\cdot\eta')\\ \eta'^A\partial'_A F(\eta\cdot\eta')&=-l_B F(\eta\cdot\eta')\end{aligned} \tag{6.2}$$

are consistent only if $l_A=l_B$ (remember that $\eta\cdot\eta'=-\tfrac{1}{2}kk'(x-x')^2$).

We thus obtain the unique solution

$$\begin{aligned}F(\eta\cdot\eta')&=C_{AB}(\eta\cdot\eta')^{-l_A} && l_A=l_B\\ &=0 && l_A\neq l_B\end{aligned} \tag{6.3}$$

or on space-time

$$\begin{aligned}F((x-x')^2)&=C'_{AB}[(x-x')^{-2}]^{-l_A} && l_A=l_B\\ &=0 && l_A\neq l_B\,.\end{aligned} \tag{6.4}$$

The generalization of the selection rule (6.3) to the two-point function of two irreducible tensor fields is given by the following: "V.E.V. selection rule": The V.E.V.

$$\langle 0|\Psi_{A_1\ldots A_n}(\eta)\,\Phi_{B_1\ldots B_m}(\eta')|0\rangle=F_{A_1\ldots A_n B_1\ldots B_m}(\eta,\eta') \tag{6.5}$$

is non-vanishing if and only if $l_A = l_B$ and $n = m$, where $-l_A - l_B$ are the degrees of homogeneity of the two tensor fields. Proof:

The scalar function

$$F(\eta, \eta') = \eta'^{A_1} \dots \eta'^{A_n} \eta^{B_1} \dots \eta^{B_m} F_{A_1 \dots A_n B_1 \dots B_m}(\eta, \eta') \tag{6.6}$$

is homogeneous of degree $-l_A + m$, $-l_B + n$ in k and k' respectively, so (6.6) is consistent only for [65]

$$l_A - l_B = m - n \,. \tag{6.7}$$

The scalar (6.6) is the contribution to the V.E.V. (6.5) of the covariant

$$\eta_{A_1} \dots \eta_{A_n} \eta'_{B_1} \dots \eta'_{B_m} (\eta \cdot \eta')^{-l_A - n} \tag{6.8}$$

which, by itself, satisfies the trace and transversality condition (5.15c) with fixed homogeneity degree for l_A, l_B related by (6.7). The whole set of allowed covariants is obtained from (6.8) by performing one of the following operations

1) Permutations of η and η',
2) substitution of an arbitrary number of couples (η, η') with g_{AB},
3) substitution of an equal number of (η, η) and (η', η') couples with a corresponding number of symbols g_{AB}.

This concludes the proof: in fact, we can transform into space-time, using Eq. (5.12):

$$\begin{aligned} &\langle 0 | O_{\alpha_1 \dots \alpha_n}(x) \, O_{\beta_1 \dots \beta_m}(x') | 0 \rangle \\ &= k^{l_A} k'^{l_B} (e^{-ix\pi})^{A_1 \dots A_n}_{\alpha_1 \dots \alpha_n} (e^{-ix'\pi})^{B_1 \dots B_m}_{\beta_1 \dots \beta_m} \langle 0 | \Psi_{A_1 \dots A_n}(\eta) \, \Phi_{B_1 \dots B_m}(\eta') | 0 \rangle \,. \end{aligned} \tag{6.9}$$

On the other hand the transformation (6.9) on the coordinates gives in general (see Chapter 5)

$$(e^{-ix\pi})^{A_1 \dots A_n}_{\alpha_1 \dots \alpha_n} \bar{\eta}_{A_1} \dots \bar{\eta}_{A_n} = k^n (\bar{x} - x)_{\alpha_1} \dots (\bar{x} - x)_{\alpha_n}$$

so that the covariants defined above, satisfying the supplementary, traceless and symmetry conditions, do not contribute to (6.9) unless $n = m$; in fact, consider as an example the leading light-cone contribution to the V.E.V. (6.5), which comes out from the covariant

$$\eta'_{A_1} \dots \eta'_{A_n} \eta_{B_1} \dots \eta_{B_m} F'(\eta \cdot \eta') \,.$$

Using homogeneity arguments, $O(4,2)$ covariance implies $l_A - l_B = n - m$ (and then $F'(\eta \cdot \eta') \sim (\eta \cdot \eta')^{-l_B - n}$) which turns out to be consistent with Eq. (6.7) only for $m = n$ and then $l_A = l_B$.

It is interesting to derive the above theorem directly in space-time. In fact the V.E.V. (6.9) is nothing but the contribution of the identity operator to the operator product expansion

$$O_{\alpha_1\ldots\alpha_n}(x)\,O_{\beta_1\ldots\beta_m}(0)=F_{\alpha_1\ldots\alpha_n\beta_1\ldots\beta_m}(x)\,I+\cdots \tag{6.10}$$

where the dots stand for local operators which do not contribute to the V.E.V. Commuting both sides with the generator of special conformal transformation K_λ we obtain the equation

$$\left(2x_\lambda x^\mu\frac{\partial}{\partial x_\mu}-x^2\frac{\partial}{\partial x_\lambda}+2l_A x_\lambda-2ix^\nu\Sigma^{(n)}_{\lambda\nu}\right)F_{\alpha_1\ldots\alpha_n\beta_1\ldots\beta_m}(x)=0 \tag{6.11}$$

where

$$\begin{aligned}F_{\alpha_1\ldots\alpha_n\beta_1\ldots\beta_m}(x)=C_1\,x_{\alpha_1}\ldots x_{\alpha_n}x_{\beta_1}\ldots x_{\beta_m}(1/x^2)^{\frac{1}{2}(l_A+l_B+m+n)}\\ +\text{ less singular terms in }x^2\end{aligned} \tag{6.12}$$

(the strength of the singularity being fixed by dilatation covariance). Order by order in x^2 Eq. (6.11) gives a set of homogeneous relations between differently singular terms in (6.12) which turn out to be consistent only if $l_A-l_B=0$ and $n-m=0$. However, in the light-cone limit $x^2\to 0$, retaining only the most singular term in (6.12), Eq. (6.11) gives the less stringent condition

$$l_A-l_B=n-m\,. \tag{6.13}$$

This is due to the fact that conformal invariance on the light-cone is less restrictive than full conformal invariance. The light-cone limit can also be obtained in the six-dimensional formalism. Let us consider, as an example, the simple case

$$\langle 0|\Psi_A(\eta)\,C(\eta')|0\rangle=C_1\eta_A(\eta\cdot\eta')^{-l_B}+C_2\eta'_A(\eta\cdot\eta')^{-l_A}\,. \tag{6.14}$$

In the limit $\eta\cdot\eta'\to 0$ both terms on the right-hand-side satisfy the transversality condition, the first one identically, the second one to an order $O(\eta\cdot\eta')$ with respect to (6.14). On space-time we have

$$\langle 0|O_\mu(x)\,C(x')|0\rangle\underset{(x-x')^2\to 0}{=}C'_2(x-x')_\mu[(x-x')^{-2}]^{l_A}\,. \tag{6.15}$$

Note that this terms verifies (6.13), i.e. $l_A-l_B=1$.

However full conformal invariance would imply $c_2=0$ and the vanishing of (6.15).

For completeness we write down the full conformal invariant two-point function for $n=1,2$:

"Current correlation function"

$$\langle 0|j_\mu(x)\,j_\nu(0)|0\rangle=C(1/x^2)^{l_j+1}(x_\mu x_\nu-\tfrac{1}{2}x^2g_{\mu\nu})\,. \tag{6.16}$$

"Gravitational" correlation function

$$\langle 0|\theta_{\mu\nu}(x)\,\theta_{\varrho\sigma}(0)|0\rangle = C'(1/x^2)^{l_\theta+2}[4x_\mu x_\nu x_\varrho x_\sigma - x^2(x_\mu x_\sigma g_{\nu\varrho} + x_\nu x_\sigma g_{\mu\varrho} + x_\mu x_\varrho g_{\nu\sigma} + x_\nu x_\varrho g_{\mu\sigma}) + \tfrac{1}{2}x^4(g_{\mu\varrho}g_{\nu\sigma} + g_{\nu\varrho}g_{\mu\sigma} - \tfrac{1}{2}g_{\mu\nu}g_{\varrho\sigma})]\,. \tag{6.17}$$

Note that the Ward-identities are automatically satisfied for $l_J = 3$, $l_\theta = 4$.

6.2. Three Point-Functions

We now consider the general case of n-point functions (scalar fields for simplicity)

$$\langle 0|A_1(\eta_1)\ldots A_n(\eta_n)|0\rangle = F_n(\eta_1\ldots\eta_n) \tag{6.18}$$

where F_n depends on the $\frac{1}{2}n(n-1)$ scalar products $(\eta_1\cdot\eta_2)\ldots(\eta_{n-1}\cdot\eta_n)$. Moreover F_n verifies the n-constraints

$$\eta^i\partial_{A_i}F_n((\eta_1\cdot\eta_2)\ldots(\eta_{n-1}\cdot\eta_n)) = -l_{A_i}F_n((\eta_1\cdot\eta_2)\ldots(\eta_{n-1}\cdot\eta_n)) \tag{6.19}$$

so that it only depends at least on $\frac{1}{2}n(n-1)-n=\frac{1}{2}n(n-3)$ independent variables. As a consequence, the three point function is completely determined[3] and it turns out to be

$$F_3((\eta_1\cdot\eta_2),(\eta_1\cdot\eta_3),(\eta_2\cdot\eta_3)) = C_{123}(\eta_1\cdot\eta_2)^{-\frac{1}{2}(l_{A_1}+l_{A_2}-l_{A_3})}\cdot(\eta_1\cdot\eta_3)^{-\frac{1}{2}(l_{A_1}+l_{A_3}-l_{A_2})}(\eta_2\cdot\eta_3)^{-\frac{1}{2}(l_{A_2}+l_{A_3}-l_{A_1})} \tag{6.20}$$

or, on space-time,

$$\langle 0|A_1(x_1)\,A_2(x_2)\,A_3(x_3)|0\rangle = C'_{123}[(x_1-x_2)^{-2}]^{\frac{1}{2}(l_{A_1}+l_{A_2}-l_{A_3})}\cdot[(x_1-x_3)^{-2}]^{\frac{1}{2}(l_{A_1}+l_{A_3}-l_{A_2})}\cdot[(x_2-x_3)^{-2}]^{\frac{1}{2}(l_{A_1}+l_{A_3}-l_{A_1})}\,. \tag{6.21}$$

Let us now consider the two interesting cases of the vector-scalar-scalar and tensor-scalar-scalar vertex[4]. Using the same techniques sketched above one obtains

"Electromagnetic vertex"

$$\langle 0|J_\mu(x)\,A(y)\,B(0)|0\rangle = C_{JAB}[(x-y)^{-2}]^{\frac{1}{2}(l_J+l_A-l_B+1)}\cdot[1/x^2]^{\frac{1}{2}(l_J+l_B-l_A+1)}[1/y^2]^{\frac{1}{2}(l_A+l_B-l_J+1)}[x^2(x-y)_\mu-(x-y)^2x_\mu] \tag{6.22}$$

---

[3] Applications of the conformal invariant solution of the vertex function have been given by: Polyakov [138] in the framework of the physics of phase transitions and used by D'Eramo, Parisi and Peliti to obtain boostrap equations for the critical indices; and by Migdal [128] in the context of field theory.

[4] Consequences of exact conformal symmetry on the vector and axial-vector vertex functions have been investigated by Schreier [145].

"gravitational vertex"

$$\begin{aligned}\langle 0|\theta_{\mu\nu}(x)\,A(y)\,B(0)|0\rangle = C_{\theta AB}&[(x-y)^{-2}]^{\frac{1}{2}(l_\theta+l_A+l_B+2)}\\ &\cdot[1/x^2]^{\frac{1}{2}(l_\theta+l_B-l_A+2)}\cdot[1/y^2]^{\frac{1}{2}(l_A+l_B-l_\theta+2)}\\ &\cdot[x^4(x-y)_\mu(x-y)_\nu+(x-y)^4x_\mu x_\nu-x^2(x-y)^2\\ &\cdot((x-y)_\mu x_\nu+x_\mu(x-y)_\nu)-\tfrac{1}{4}x^2y^2(x-y)^2g_{\mu\nu}]\end{aligned} \tag{6.23}$$

the Ward identities associated to the vertices (6.22), (6.23) give the following selection rule

$$\langle 0|J_\mu(x)\,A(y)\,B(0)|0\rangle = 0 = \langle 0|\theta_{\mu\nu}(x)\,A(y)\,B(0)|0\rangle \tag{6.24}$$

unless $l_A = l_B$.

Eq. (6.24) is a particular case of the general selection rule valid for conserved irreducible 4-tensors[5] $O_{\alpha_1\ldots\alpha_n}(x)$ (and then $l_n = 2+n$)

$$\langle 0|O_{\alpha_1\ldots\alpha_n}(x)\,A(y)\,B(0)|0\rangle = 0 \quad \text{unless} \quad l_A = l_B\,. \tag{6.25}$$

7. Operator Products and Conformal Invariance on the Light-Cone

7.1. Operator Product Expansions on the Light-Cone

We want to discuss the behaviour under conformal transformations of the operator expansion of the product of two local operators $A(x)$, $B(x')$ in the light-cone limit $(x-x')^2\to 0$. The possible relevance of conformal symmetry on this limit has already been discussed in the introduction. We again stress that the hypersurface $(x-x')^2=0$ is invariant under the action of the conformal group on space-time. The operator expansion previously discussed in the introduction can be regarded, from an algebraic point of view, as a decomposition of a direct product of two tensors A and B into irreducible tensor representations [64, 65]. The classifications and the properties [67] of these representations have been discussed in Chapter 5. In particular, the contribution of the identical representation to the operator product has been examined in Chapter 6.

The light-cone limit, $(x-x')^2\to 0$, of the product $A(x)\,B(x')$ corresponds to the (conformally) covariant limit

$$A(\eta)\,B(\eta') \quad \text{for} \quad \eta\cdot\eta'\to 0\,(\eta\cdot\eta' = -\tfrac{1}{2}kk'(x-x')^2)\,.$$

We assume $A(x)$, $B(x')$ to be conformal scalars, i.e. of definite dimensions l_A, l_B and satisfying

$$[A(0), K_\lambda] = [B(0), K_\lambda] = 0\,.$$

[5] The selection rule (6.25) in the particular case $n=1$, has been derived independently by Migdal.

The most general decomposition into irreducible tensor operators is of the form

$$A(\eta)\,B(n')\underset{\eta\cdot\eta'\to 0}{=}\sum_{n=0}^{\infty} E_n(\eta\cdot\eta')\,D^{(n)A_1\ldots A_n}(\eta,\eta')\,\Psi_{A_1\ldots A_n}(\eta') \tag{7.1}$$

where

$$E_n(\eta\cdot\eta') = C_n(\eta\cdot\eta')^{-\frac{1}{2}(l_A+l_B-l_n+n)} \tag{7.2}$$

with C_n constant and $-l_n$ being the homogeneity degree of $\Psi_{A_1\ldots A_n}(\eta')$. The general structure of $D^{(n)A_1\ldots A_n}(\eta,\eta')$ is

$$D^{(n)A_1\ldots A_n}(\eta,\eta') = \sum_{m=0}^{n} \eta^{A_1}\ldots\eta^{A_{n-m}}\eta'^{A_{n-m+1}}\ldots\eta'^{A_n}D^{(n,m)}(\eta,\eta')\,C_{nm} \tag{7.3}$$

where C_{nm} are constants and $D^{(n,m)}(\eta,\eta')$ is a differential operator defined on the cones $\eta^2=\eta'=0$, finite for $\eta\cdot\eta'=0$. These conditions are equivalent to the requirement that $D^{(n,m)}(\eta,\eta')$ be homogeneous in k/k' (in the parametrization introduced in Chapter 5). Covariance of the expansion under the group of dilatations $\eta_A\to\lambda\eta_A$ on the six-dimensional cones $\eta^2=\eta'^2=0$ implies (7.2) and fixes the homogeneity degree of $D^{(n,m)}(\eta,\eta')$ to be $h=-\frac{1}{2}(l_A-l_B+l_n+n)$. It is shown in the next section that $D^{(n,m)}(\eta,\eta')$ is uniquely determined, and for $\eta\cdot\eta'=0$ it reduces to [66]

$$D^{(n,m)}(\eta,\eta') = (\eta\cdot\partial')^h = (\eta^A\,\partial/\partial\eta'_A)^h \tag{7.4}$$

where by $(\eta\cdot\partial')^h$ we mean the application of $\eta\partial'$ h times, when h is integer, and its analytical continuation for non-integer h (as specified later). By simple algebraic steps one can recognize that the $n+1$ tensor covariants in Eq. (7.3) are all proportional and therefore one can consistently take

$$D^{(n)A_1\ldots A_n}(\eta,\eta') = \eta^{A_1}\ldots\eta^{A_n}(\eta\cdot\partial')^{-\frac{1}{2}(l_A-l_B+l_n+n)} \tag{7.5}$$

to define the application of $(\eta\cdot\partial')^{-\frac{1}{2}(l_A-l_B+l_n+n)}$ on $\Psi_{A_1\ldots A_n}(\eta')$ we write

$$\eta\cdot\partial' = (k/k')\,(k'\,\partial/\partial k' + (x-x')\cdot\partial') \quad \text{at} \quad (x-x')^2=0\,. \tag{7.6}$$

One has the following property, valid for β integer

$$\begin{aligned}(\eta\cdot\partial')^h &= (-k/k')\,\frac{\Gamma(l_n+\beta)}{\Gamma(l_n)}\sum_{J=0}^{\beta}\frac{\Gamma(l_n)}{\Gamma(l_n+J)}(-1)^J(x-x')^{\alpha_1}\ldots(x-x')^{\alpha_J}\\ &\cdot\partial'_{\alpha_1}\ldots\partial'_{\alpha_J} = (-k/k')\,\frac{\Gamma(l_n+\beta)}{\Gamma(l_n)}\;{}_1F_1(-\beta;l_n;(x-x')\,\partial') \quad (\partial'_\alpha=\partial/\partial x'_\alpha)\end{aligned} \tag{7.7}$$

where ${}_1F_1$ is the confluent hypergeometric function [31] and $[(x-x')\partial']^J$ stands for $(x-x')^{\alpha_1}\ldots(x-x')^{\alpha_J}\,\partial'_{\alpha_1}\ldots\partial'_{\alpha_J}$.

Eq. (7.7) defines $(\eta\cdot\partial')^\beta$ also for β non-integer. Making use of Eqs. (7.5) and (7.7), Eq. (7.1) can be rewritten as

$$A(x)\,B(x') \underset{(x-x')^2\to 0}{=} \sum_{n=0}^{\infty} [(x-x')^{-2}]^{\frac{1}{2}(l_A+l_B+n-l_n)} C_n^{AB} X^{A_1}\dots X^{A_n} \cdot {}_1F_1(\tfrac{1}{2}(l_A-l_B+l_n+n);\, l_n;\, (x-x')\cdot\partial')\, \tilde{\Psi}_{A_1\dots A_n}(x') \tag{7.8}$$

where

$$\Psi_{A_1\dots A_n}(x') = (k')^{l_n}\,\Psi_{A_1\dots A_n}(\eta'), \qquad X^A = (x_\mu, (1+x^2)/2, (1-x^2)/2). \tag{7.9}$$

We note that in the limit $(x-x')^2\to 0$ one can consistently put

$$\tilde{\Psi}_{\alpha_1\dots\alpha_J\times\times\dots}(0) = 0 \tag{7.10}$$

where the crosses $\times\times\times\dots$ stand for indices 5 or 6. In fact such components of $\tilde{\Psi}$ correspond to less dominant contributions, of order $[(x-x')^2]^{n-J}$ with respect to the leading singularity of the particular tensor representation [67].

Making use of the fundamental integral representation, [4]

$${}_1F_1(a;c;z) = \frac{\Gamma(c)}{\Gamma(a)\,\Gamma(c-a)} \int_0^1 du\cdot u^{a-1}(1-u)^{c-a-1} e^{uz} \tag{7.11}$$

of the identity

$$e^{-ix\cdot\pi}\partial_\mu e^{ix\cdot\pi} = \partial_\mu + i\pi_\mu \tag{7.12}$$

and of the definition

$$\tilde{\Psi}_{A_1\dots A_n}(x') = (e^{ix'\pi} O)_{A_1\dots A_n}(x') \tag{7.13}$$

one has

$$A(x)\,B(x') \underset{(x-x')^2\to 0}{=} \sum_{n=0}^{\infty} [(x-x')^{-2}]^{\frac{1}{2}(l_A+l_B+n-l_n)} C_n^{AB} X^{A_1}\dots X^{A_n} e^{ix'\cdot\pi} \cdot {}_1F_1(\tfrac{1}{2}(l_A-l_B+n+l_n);\, l_n;(x-x')\,(\partial'+i\pi))\, O_{A_1\dots A_n}(x'). \tag{7.14}$$

Taking $x'=0$ and inserting Eq. (7.11) one finds

$$\begin{aligned} X^{A_1}\dots X^{A_n}\,{}_1F_1(a;c;x(\partial+i\pi))\,O_{A_1\dots A_n}(0) &= X^{A_1}\dots X^{A_n}\frac{\Gamma(c)}{\Gamma(a)\,\Gamma(c-a)} \\ &\quad\cdot\int_0^1 du\, u^{a-1}(1-u)^{c-a-1} e^{iux\cdot\pi} O_{A_1\dots A_n}(ux) \\ &= X^{A_1}\dots X^{A_n}\frac{\Gamma(c)}{\Gamma(a)\,\Gamma(c-a)}\int_0^1 du\, u^{a-1}(1-u)^{c-a-1}\tilde{\Psi}_{A_1\dots A_n}(ux). \end{aligned} \tag{7.15}$$

From the property (6.10) we are able to write

$$\begin{aligned} X^{A_1} \dots X^{A_n} \tilde{\Psi}_{A_1 \dots A_n}(ux) &= (x-ux)^{\alpha_1} \dots (x-ux)^{\alpha_n} O_{\alpha_1 \dots \alpha_n}(ux) \\ &= (1-u)^n x^{\alpha_1} \dots x^{\alpha_n} O_{\alpha_1 \dots \alpha_n}(ux) \end{aligned} \tag{7.16}$$

and we finally obtain the light-cone expansion formula [64, 65]

$$\begin{aligned} A(x)\,B(0) \underset{x^2 \to 0}{=} \sum_{n=0}^{\infty} \tilde{C}_n^{AB} (1/x^2)^{\frac{1}{2}(l_A + l_B + n - l_n)} x^{\alpha_1} \dots x^{\alpha_n} \\ \cdot {}_1F_1\left(\tfrac{1}{2}(l_A - l_B + l_n + n); l_n + n; x \cdot \partial\right) O_{\alpha_1 \dots \alpha_n}(0) \end{aligned} \tag{7.16'}$$

where $\tilde{C}_n^{AB}$ are unknown constants.

It is interesting to observe that the result (7.16) could be obtained by application of the Jacobi identities of conformal algebra on the expansion

$$A(x)\,B(0) \underset{x^2 \to 0}{=} \sum_{n=0}^{\infty} (1/x^2)^{\frac{1}{2}(l_A + l_B + n - l_n)} \sum_{m=n}^{\infty} C_{nm}^{AB} x^{\alpha_1} \dots x^{\alpha_m} O_{\alpha_1 \dots \alpha_m}^{nm}(0) \tag{7.17}$$

where the families of hermitean operators

$$O_{\alpha_1 \dots \alpha_m}^{nm}(0) = [\dots [O_{\alpha_1 \dots \alpha_m}^{(n)}(0), P_{\alpha_{n+1}}] \dots P_{\alpha_m}]\,(i)^{m-n} \tag{7.18}$$

together contribute a complete set for expansion on the light-cone. Each family (i.e., fixed n) transforms irreducibly under the conformal algebra in space-time [63]. $O_{\alpha_1 \dots \alpha_n}^{(n)}(0)$ is the lowest dimension operator, in the representation, for which $K_\lambda = 0$. Note that P_λ and K_λ are dimension-rising and -lowering operators respectively within each irreducible representation. Commuting both sides of Eq. (7.17) with the generators of special conformal transformation K_λ one gets:

$$\begin{aligned} -i[A(x)\,B(0), K_\lambda] &= \sum_{n=0}^{\infty} (1/x^2)^{\frac{1}{2}(l_A + l_B + n - l_n)} \sum_{m=n}^{\infty} C_{nm}^{AB} \\ &\cdot 2[l_A + m - \tfrac{1}{2}(l_A + l_B + n - l_n)]\, x_\lambda x^{\alpha_1} \dots x^{\alpha_n} - x^2 \\ &\cdot \sum_{i=1}^{m} g_\lambda^{\alpha_i} x^{\alpha_1} \dots x^{\hat{\alpha}_i} \dots x^{\alpha_n} O_{\alpha_1 \dots \alpha_m}^{nm}(0) = \sum_{n=0}^{\infty} (1/x^2)^{\frac{1}{2}(l_A + l_B + n - l_n)} \\ &\cdot \sum_{m=n}^{\infty} C_{nm}^{AB} x^{\alpha_1} \dots x^{\alpha_m} [O_{\alpha_1 \dots \alpha_m}^{nm}(0), K_\lambda]\,(-i) \end{aligned} \tag{7.18'}$$

where:

$$\begin{aligned} i[O_{\alpha_1 \dots \alpha_m}^{nm}(0), K_\lambda] &= -(m-n)(2l_n + m - n - 1) \underset{\{\alpha\}}{S} g_{\alpha_m \lambda} O_{\alpha_1 \dots \alpha_n, \alpha_{n+1} \dots \alpha_{m-1}}^{nm-1}(0) \\ &- (m-n)(m-n-1) \underset{\{\alpha\}}{S} (g_{\alpha_m \lambda} O_{\alpha_1 \dots \alpha_n, \alpha_{n+1} \dots \alpha_{m-1}}^{nm-1}(0) \\ & g_{\alpha_{m-1} \alpha_m} O_{\alpha_1 \dots \alpha_n, \alpha_{n+1} \dots \alpha_{m-1}}^{nm-1}(0) + 2(m-n) \\ &\cdot \underset{\{\alpha\}}{S} (g_{\alpha_1 \alpha_m} O_{\alpha_2 \dots \alpha_m \lambda, \alpha_{n+1} \dots \alpha_{m-1}}^{nm-1}(0) - g_{\lambda \alpha_1} O_{\alpha_2 \dots \alpha_n \alpha_m, \alpha_{n+1} \dots \alpha_{m-1}}^{nm-1}(0)). \end{aligned} \tag{7.19}$$

From the structure of the commutator given by Eq. (7.18) it is manifest that the leading terms on the light-cone, which are proportional to x_λ, arise from those terms in the commutator (7.19) which are proportional to $g_\lambda^{\alpha_i}$. Their symmetrization with respect to the $\{\alpha\}$ indices gives

$$-2(m-n)(l_n+m-1)\underset{\{\alpha_i\}}{S}\, g_{\alpha_i\lambda}\, O^{nm-1}_{\alpha_1\ldots\hat{\alpha}_i\ldots\alpha_m}(0). \tag{7.20}$$

Inserting Eq. (7.20) in the commutator (7.18) we obtain

$$\begin{aligned} -i[A(x)\,B(0), K_\lambda] &= \sum_{n=0}^{\infty} (1/x^2)^{\frac{1}{2}(l_A+l_B+n-l_n)} \sum_{m=n}^{\infty} C^{AB}_{nm} \\ &\quad \cdot 2[l_A+m-\tfrac{1}{2}(l_A+l_B+n-l_n)]\, x_\lambda x^{\alpha_1}\ldots x^{\alpha_m} O^{nm}_{\alpha_1\ldots\alpha_m}(0) \\ &= \sum_{m=0}^{\infty} (1/x^2)^{\frac{1}{2}(l_A+l_B+n-l_n)} \sum_{m=1}^{\infty} C^{AB}_{nm}\, b(n,m)\, x_\lambda x^{\alpha_1}\ldots x^{\alpha_{m-1}} O^{nm-1}_{\alpha_1\ldots\alpha_{m-1}}(0) \end{aligned} \tag{7.21}$$

where $b(n,m)=(m-n)\,(l_n+m)$.

Eq. (7.21), by comparing equal power in x, gives rise to the recurrence relation

$$C^{AB}_{n,n+k-1}[\tfrac{1}{2}(l_A-l_B+l_n+n)+k-1]=C^{AB}_{n,n+k}\,k(l_n+n+k) \tag{7.22}$$

giving the solution

$$C^{AB}_{n,n+k}=\frac{\Gamma(\frac{1}{2}(l_A-l_B+l_n+n)+k)\,\Gamma(l_n+n)}{k!\,\Gamma(\frac{1}{2}(l_A-l_B+l_n+n))\,\Gamma(l_n+n+k)}\,C^{AB}_{n,n}\,. \tag{7.23}$$

Eq. (7.23) brings about, by Taylor expansion, the confluent hypergeometric function

$${}_1F_1(\tfrac{1}{2}(l_A-l_B+l_n+n);\, l_n+n;\, x\cdot\partial)=\sum_{k=0}^{\infty}\frac{C^{AB}_{n,n+k}}{C^{AB}_{n,n}}\,(x\cdot\partial)^k\,. \tag{7.24}$$

7.2. Properties of the Light-Cone Expansion

The light-cone expansion (7.16) is covariant under the whole conformal algebra of space-time. It is manifestly covariant under homogeneous Lorentz transformations and dilatations. Covariance under special conformal transformations has been checked above. It is particularly interesting to exploit how the property of translation covariance is verified [64].

Assume $A(x)$, $B(x)$, $O^{(n)}_{\alpha_1\ldots\alpha_n}(x)$ to be hermitean (the hermiticity of the basis tensors guarantees the correct causality support for the commutator of $[A(x), B(0)]$). Then, translating Eq. (7.16) by $-x$, changing x

into $-x$, and taking the hermitean conjugate we obtain

$$\begin{aligned} B(x)\,A(0) \underset{x^2\to 0}{=} \sum_{n=0}^{\infty} (1/x^2)^{\frac{1}{2}(l_A+l_B+n-l_n)} C_n^{AB}(-1)^n x^{\alpha_1} \dots x^{\alpha_n} \\ \cdot {}_1F_1(\tfrac{1}{2}(l_A-l_B+l_n+n); l_n+n; -x\cdot\partial)\, O^{(n)}_{\alpha_1\dots\alpha_n}(x)\,. \end{aligned} \tag{7.25}$$

However, using the Kummer transformation [5]

$$ {}_1F_1(a;c;z) = {}_1F_1(c-a;c;-z)\,\mathrm{e}^z \tag{7.26}$$

we have

$$\begin{aligned} &{}_1F_1(\tfrac{1}{2}(l_A-l_B+l_n+n); l_n+n; -x\cdot\partial)\, O^n_{\alpha_1\dots\alpha_n}(x) \\ &\quad = {}_1F_1(\tfrac{1}{2}(l_B-l_A+l_n+n); l_n+n; x\cdot\partial)\, O^n_{\alpha_1\dots\alpha_n}(0)\,. \end{aligned} \tag{7.27}$$

Thus one recovers the expansion (7.16) with the interchange $A\leftrightarrow B$, provided $C_n^{AB}=(-1)^n C_n^{BA}$ (hermiticity property) [64]. We note that for $A=B$ it follows that $C_n^{AA}=0$ for n odd, which is nothing but the well known property that $\langle p|[A(x),A(0)]|p\rangle$ is an even function in x. In this case Eq. (7.16) simplifies to

$$\begin{aligned} A(x)\,A(0) = \sum_{n=\text{even}} C_n (1/x^2)^{\frac{1}{2}(2l_A+n-l_n)} x^{\alpha_1} \dots x^{\alpha_n} \\ \cdot (x\cdot\partial)^{\frac{1}{2}(1-l_n+n)} I_{\frac{1}{2}(l_n+n-1)}(x\cdot\partial/2)\, O_{\alpha_1\dots\alpha_n}(0) \end{aligned} \tag{7.28}$$

where $I_\nu(z)$ is a modified Bessel function [6].

As a final point we remark that, when $\frac{1}{2}(l_A-l_B+n+l_n)=-h$ (h non-negative integer) the contribution of the corresponding representation to the expansion is [7]

$$(1/x^2)^{\frac{1}{2}(l_A+l_B+n-l_n)} x^{\alpha_1} \dots x^{\alpha_n} L_h^{(l_n+n-1)}(x\cdot\partial)\, O_{\alpha_1\dots\alpha_n}(0) \tag{7.29}$$

where $L_h^{(l_n+n-1)}(x\cdot\partial)$ is a Laguerre polynomial of order h, so that only a finite number of terms in the representation contributes to the expansion.

7.3. Causality on the Light-Cone

A remarkable property of the conformal covariant expansion given by Eq. (7.16) is its deep connection to the causality requirement [64]. This requirement amounts to the vanishing of the commutator $[A(x)B(0), C(y)]$ (for any local operator $C(y)$) for $y^2<0$ and $(x-y)^2<0$; in particular, for $x^2=0$ the commutator vanishes for $y^2<2xy$.

[6] ${}_1F_1(a;2a;z)=\Gamma(a+\frac{1}{2})\,(z/4)^{1/2-a}\,\mathrm{e}^{z/2}\,I_{a-1/2}(z/2)$ [4].

Writing the conformal covariant expansion in the integral form

$$A(x)\,B(0) \underset{x^2\to 0}{=} \sum_n C_n^{AB}(1/x^2)^{\frac{1}{2}(l_A+l_B+n-l_n)}\,x^{\alpha_1}\ldots x^{\alpha_n} \cdot \int_0^1 du\, u^{\frac{1}{2}(l_A-l_B+l_n+n)-1}(1-u)^{\frac{1}{2}(l_B-l_A+l_n+n)-1}\,O^n_{\alpha_1\ldots\alpha_n}(ux) \tag{7.30}$$

each irreducible component satisfies the causality property provided

$$[O^n_{\alpha_1\ldots\alpha_n}(ux), C(y)]=0 \quad \text{if} \quad (ux-y)^2 = y^2 - 2ux\cdot y < 0\,. \tag{7.31}$$

But $y^2 < 2uxy$ holds for $0 \leqq u \leqq 1$ if both $y^2<0$ and $y^2<2x\cdot y$ and viceversa. These considerations imply that causality does not impose any constraint on the sum appearing in (7.27) as it is independently satisfied by each contributing irreducible representation.

In Eq. (7.27) the functions

$$f_n^{AB}(u) = u^{\frac{1}{2}(l_A-l_B+l_n+n)-1}(1-u)^{\frac{1}{2}(l_B-l_A+l_n+n)-1} \tag{7.32}$$

are nothing but the Clebsch-Gordon coefficients for the decomposition of the product of two irreducible representations into irreducible components on a continuous basis (see Appendix D), whereas the coefficients $C^{AB}_{n,n+k}$ (see Eq. (7.23)) refer to a discrete basis. The relation between the two sets turns out to be

$$\int_0^1 du\, u^k f_n^{AB}(u) = \frac{\Gamma(\frac{1}{2}(l_A-l_B+l_n+n))\,\Gamma(\frac{1}{2}(l_B-l_A+l_n+n))}{k!\,\Gamma(l_n+n)}\;\frac{C^{AB}_{n,n+k}}{C^{AB}_{n,n}}\,. \tag{7.33}$$

Eq. (7.33) gives the coefficients C^{AB}_{nm} as momenta, integrated over a compact support, of the functions $f_n^{AB}(u)$. In the continuous basis it also turns out to be straightforward to obtain the most singular contribution on the light-cone to the product of two arbitrary tensor operators, irreducible under the conformal algebra. One obtains, exactly as before

$$J^A_{\alpha_1\ldots\alpha_{n_A}}(x)\,J^B_{\beta_1\ldots\beta_{n_B}}(0) = \sum_{n=0}^{\infty}(1/x^2)^{\frac{1}{2}(l_A+l_B-l_n+n_A+n_B+n)}C_n^{AB} \cdot x_{\alpha_1}\ldots x_{\alpha_{n_A}}\,x_{\beta_1}\ldots x_{\beta_{n_B}}\,x^{\mu_1}\ldots x^{\mu_n} \cdot {}_1F_1(\tfrac{1}{2}(l_A-l_B+l_n+n+n_B-n_A);\, l_n+n;\, x\cdot\partial)\,O^n_{\mu_1\ldots\mu_n}(0)\,. \tag{7.34}$$

A connection holds between the terms appearing in the conformal covariant light-cone expansion and the three-point functions described in Chapter 6. The connection which will be again considered later on in the frame of full conformal covariance, is discussed in Appendix E.

8. Consequences of Exact Conformal Symmetry on Operator Product Expansions

8.1. Operator Expansions on Space-Time

In this chapter we study the consequences of the application of strict conformal symmetry on operator products [66]. Note that this requirement is more stringent than conformal invariance on the light-cone. In fact, as one can easily verify, the application of the generators of the special conformal transformations, K_λ, gives rise to a set of equations which relate leading terms to the non-leading ones on the light-cone.

This implies, in particular, that exact conformal symmetry gives the complete structure of the contribution of each irreducible representation to the operator product expansion, at each order of x^2, adding, in a well-definite way, a whole (infinite) set of less dominant powers of x^2 to the dominant singularity on the light-cone.

Moreover we would like to stress that full conformal invariance on operator expansions, even if it is a very stringent assumption, could be useful in the investigation of the properties of the so called skeleton theories. Motivated by the previous discussion we now consider the manifestly covariant expansion (7.1) of the previous section, but not limiting to $\eta \cdot \eta' \to 0$. We immediately see that the expansion has the general structure given by Eq. (7.3) of the previous section, provided we modify the definition of the "orbital" operator which appears on Eq. (7.4). In fact we observe that the operator

$$\eta \cdot \partial' = (k/k')\,[(x-x') \cdot \partial' + k'\,\partial/\partial k'] + 2(\eta \cdot \eta')\,\partial/\partial \eta'^2 \tag{8.1}$$

is not defined on the cone $\eta'^2 = 0$ for $\eta \cdot \eta' \neq 0$.

However, as

$$\Box_6' = (1/k'^2)\,\Box_4' + 4(1 + k'\,\partial/\partial k')\,\partial/\partial \eta'^2 \tag{8.2}$$

we have that the new operator

$$D(\eta, \eta') = \eta \cdot \eta' \,\Box_6' - 2\eta \cdot \partial'(1 + \eta' \cdot \partial') \tag{8.3}$$

is a differential operator, defined on the cone $\eta'^2 = 0$ for $\eta \cdot \eta' \neq 0$ and which essentially reduces to $\eta \cdot \partial'$ for $\eta \cdot \eta' = 0$. It is also evident that, apart from a multiplicative constant, this is the unique operator defined on $\eta'^2 = 0$ and of homogeneity degree k/k'. Note that it can also be written in a form which is manifestly defined on the cone $\eta'^2 = 0$:

$$D(\eta, \eta') = (\eta \cdot \eta')^{-1} \eta^A \eta^C g^{BD} L'_{AB} L'_{CD} \tag{8.4}$$

where we recall that $L'_{AB} = i(\eta'_A \partial'_B - \eta'_B \partial'_A)$.

Then we have, instead of Eq. (7.4)

$$D^{(n,m)}(\eta, \eta') = D^h(\eta, \eta')$$

where h is the same as defined in the previous section. We now mention two preliminary Lemmas

I) The $n+1$ covariants defined in Eq. (7.3) are all proportional: In fact from the supplementary conditions

$$\eta^{A_1}\Psi_{A_1A_2\dots A_n}(\eta)=0 \qquad \partial^{A_1}\Psi_{A_1A_2\dots A_n}(\eta)=0$$

one deduces that

$$\eta'^A D^h(\eta,\eta')\,\Psi_{A_1\dots A_n}(\eta')=\eta^A 2h(2-h+\lambda_n)\,D^{h-1}(\eta,\eta')\,\Psi_{A_1\dots A_n}(\eta'). \tag{8.6}$$

In fact we have for $m=1$

$$\begin{aligned}&\eta'^{A_1}D(\eta,\eta')\,\Psi_{A_1\dots A_n}(\eta')\\ &\quad=\{D(\eta,\eta')\,\eta'^A+2(\eta\cdot\partial')\,\eta'^A-2(\eta\cdot\eta')\,\partial'^A+2\eta^{A_1}(1+k'\partial/\partial k')\}\\ &\quad\cdot\Psi_{A_1\dots A_n}(\eta')=2\eta^{A_1}(1-l_n)\,\Psi_{A_1\dots A_n}(\eta').\end{aligned} \tag{8.7}$$

The proof can be extended to $m=2$. Moreover, assuming its validity for arbitrary m, with simple but tedious algebra, it can be proven for $m+1$.

II) One has:

$$D^h(\eta,\eta')=(L-1)^h\sum_{J=0}^{h}\binom{h}{\Lambda}\frac{(\eta\cdot\eta')^J(2\eta\cdot\partial')^{h-J}\square_6'^J}{(L-1)^J} \tag{8.8}$$

where

$$L=-k'\partial/\partial k'$$

and

$$(L-1)^J=(l_n-1)(l_n-1+1)\dots(l_n-1+\Lambda-1)=\Gamma(l_n-1+\Lambda)/\Gamma(l_n-1).$$

Proof. For $h=1,2$ one has

$$\begin{aligned}D(\eta,\eta')&=(1-L)\,[\eta\cdot\eta'\,\square_6'/(1-L)-2(\eta\cdot\partial')]\\ &=(1-l_n)\,[\eta\cdot\eta'\,\square_6'/(1-l_n)-2(\eta\cdot\partial')],\end{aligned} \tag{8.9}$$

$$\begin{aligned}D^2(\eta,\eta')&=(1-L)^2\left[\frac{\eta\cdot\eta'\,\square_6'}{1-L}-2\eta\cdot\partial'\right]\left[\frac{\eta\cdot\eta'\,\square_6'}{1-L}-2\eta\cdot\partial'\right]\\ &=(1-l_n)(1-l_n-1)\left[\frac{2(\eta\cdot\eta')(\eta\cdot\partial')\,\square_6'}{1-l_n}+4(\eta\cdot\partial')^2+\frac{(\eta\cdot\eta')^2\,\square_6'^2}{(1-l_n)(1-l_n-1)}\right]\end{aligned} \tag{8.10}$$

which again coincides with (8.8) for $h=2$. We have used the relations

$$\begin{aligned}&(1-L)^n=(1-l_n)(1-l_n-1)\dots(1-l_n-n+1)\\ &(\eta\cdot\partial')(\eta\cdot\eta')=(\eta'\cdot\eta)(\eta\cdot\partial')\\ &\square_6'(\eta\cdot\eta')\,\square_6'=2(\eta\cdot\partial')\,\square_6'+(\eta\cdot\eta')\,\square_6'^2.\end{aligned} \tag{8.11}$$

From the Newton binomial formula for the n-th power it is possible, by a rather lengthy but trivial calculation, to show that Eq. (8.8) is also valid for the power $n+1$.

In terms of the variables (x_μ, k), and using the property given by Eq. (7.7), it turns out that

$$D^h(\eta,\eta')=(k/k')^h(-2)^h\,\frac{\Gamma(l_n-1+h)}{\Gamma(l_n-1)}\sum_{J=0}^{\infty}\frac{1}{\Lambda!}\,\frac{\Gamma(-h+\Lambda)}{\Gamma(-h)}\,\frac{\Gamma(l_n-1)}{\Gamma(l_n-1+h)}$$
$$\cdot\frac{\Gamma(l_n+h+\Lambda)}{\Gamma(l_n+2\Lambda)}\left[-\left(\frac{x-x_1}{2}\right)^2\right]^J{}_1F_1(J\cdot h;l_n+2\Lambda;(x-x')\cdot\partial')\,\Box'^J\,. \tag{8.12}$$

Eq. (8.12) certainly holds for h integer, but it can be analytically continued to non integer h.

Using the known integral representation for the ${}_1F_1$ function, Eq. (8.12) can be rewritten in the more compact form

$$D^h(\eta,\eta')=(k/k')^h(-2)^h\,\frac{\Gamma(l_n-1+h)}{\Gamma(-h)\,\Gamma(l_n-1)}\int_0^1 du\,u^{-h-1}(1-u)^{l_n+h-1}$$
$$\cdot\,e^{u(x-x')\partial'}\,{}_0F_1(l_n-1;\,-(\tfrac{1}{2}(x-x'))^2\,\Box' u(1-u))\,. \tag{8.13}$$

Eq. (8.13) is meant as a formal differential expression with all derivatives located at the right and acting to the right. Using Eq. (8.13) the general conformal covariant expansion turns out to be

$$A(x)\,B(x')=\sum_{n=0}^{\infty}[(x-x')^{-2}]^{\frac{1}{2}(l_A+l_B+n-l_n)}C_n^{AB}\int_0^1 du\,u^{\frac{1}{2}(l_A-l_B+l_n+n)-1}$$
$$\cdot(1-u)^{\frac{1}{2}(l_B-l_A+l_n-n)-1}\,e^{u(x-x')\partial'}\;{}_0F_1(l_n-1;\,-\tfrac{1}{2}(x-x')^2\,\Box' u(1-u))$$
$$\cdot\,x^{A_1}\dots x^{A_n}\Psi_{A_1\dots A_n}(x) \tag{8.14}$$

where

$$\eta_A=kx_A \quad\text{and}\quad \Psi_{A_1\dots A_n}(x)=k^{l_n}\Psi_{A_1\dots A_n}(\eta)\,.$$

From the results of Section 5 (see Eq. (5.44)) the covariant product appearing in (5.14) becomes

$$x^{A_1}\dots x^{A_n}\Psi_{A_1\dots A_n}(x')=\sum_{J=0}^{n}\binom{n}{\Lambda}\frac{\Gamma(l_n-2-n)}{\Gamma(l_n-2-n+\Lambda)}\,[-\tfrac{1}{2}(x-x')^2]^J$$
$$\cdot(x-x')^{\alpha_1}\dots(x-x')^{\alpha_{n-J}}\,\partial'_{\mu_1}\dots\partial'_{\mu_J}\,O^{\mu_1\dots\mu_J}_{\alpha_1\dots\alpha_{n-J}}(x')\,. \tag{8.15}$$

Finally, taking $x'=0$, from Eq. (8.10) one obtains

$$A(x)\,B(0)=\sum_{n=0}^{\infty}(1/x^2)^{\frac{1}{2}(l_A+l_B+n-l_n)}\,C_n^{AB}\int_0^1 du\,u^{\frac{1}{2}(l_A-l_B+l_n+n)-1}$$
$$\cdot(1-u)^{\frac{1}{2}(l_A-l_B+l_n-n)-1}\,x_u^{A_1}\ldots x_u^{A_n} \tag{8.16}$$
$$\cdot\left[{}_0F_1(l_n-1;-\tfrac{1}{4}x^2u(1-u)\,(u^{-2}\Box+2iu^{-1}\pi\cdot\partial-\pi^2))\,O\right]_{A_1\ldots A_n}(ux)$$

where

$$x_{u_A}\equiv\{(1-u)\,x_{\mu_1}\,\tfrac{1}{2}[1+(1-u)^2\,x^2],\,\tfrac{1}{2}[1-(1-u)^2\,x^2]\}$$

and the tensor $O_{A_1\ldots A_n}(x)$ has components given by Eq. (5.44).

8.2. Properties of the Strictly Conformal Covariant Operator Expansion

The expansion (8.14) is covariant under the full conformal algebra of space-time, as it is evident from its derivation. The resulting expansion, given by Eq. (8.14), is particularly interesting since it reduces automatically to the light-cone expansion (7.16) for $x^2=0$, where the ${}_0F_1$ function takes the value ${}_0F_1(l_n-1;0)=1$, and one can write $x^{A_1}\ldots x^{A_n}\,\Psi_{A_1\ldots A_n}(x')$ $=(x-x')^{\alpha_1}\ldots(x-x')^{\alpha_n}\,O_{\alpha_1\ldots\alpha_n}(x')$. Non leading terms in x^2 can simply be obtained by expanding ${}_0F_1$ in power series of x^2.

We now come to an interesting selection rule, which one obtains by noting that Eq. (8.14) is in general unacceptable for $l_n=2+n$, i.e. for "canonical dimensions" of the non-scalar representations [66].

This is a consequence of the results given in Chapter 6. In fact, we recall that, for $l_n=2+n$, the components $\psi_{\alpha_1\ldots\alpha_J xx\ldots}(0)$ ($x=5$ or 6) of the covariant tensors appearing in Eq. (8.14) are not determined from the supplementary conditions. So they remain as non-physical components and they must not be present in the expansion. However, it can be shown, using arguments of translation invariance[7], that these components cancel out in the particular case $l_A=l_B$. In this case the expansion turns out to be, for $l_n=2+n$

$$A(x)\,A(0)=\sum_{n=0}^{\infty}(1/x^2)^{l_A-1}\,C_n\int_0^1 du[u(1-u)]^n\,e^{ux\cdot\partial}$$
$$\cdot\,{}_0F_1(n+1;-\tfrac{1}{4}x^2\,\Box\,u(1-u))\,x^{\alpha_1}\ldots x^{\alpha_n}\,O^n_{\alpha_1\ldots\alpha_n}(0)\,. \tag{8.17}$$

[7] The expansions for $A(x)\,B(0)$ and $B(x)\,A(0)$ can be correlated by a sequence of translation of $-x$, reflection $x\to -x$, and hermitean conjugation. Then $l_A\leftrightarrow l_B$. However non leading terms proportional to divergences have under such sequence of operations behaviour opposite to that of the terms proportional to $x\cdot\partial$. They must therefore contain a factor (l_A-l_B).

We observe that, in this exceptional case, the terms containing divergences of the tensor $O^{(n)}_{\alpha_1\ldots\alpha_n}(0)$ add up as independent contributions in the operator product expansion [66]. These terms correspond in fact to irreducible representations in Eq. (8.14), originated by the tensor $\partial^A \psi_{AA_1\ldots A_{n-1}}(\eta)$ (which has dimension $l_n = 4+n$), according to the discussion of Chapter 6. Their contribution to the expansion (8.14) is obtained by inserting in Eq. (8.15) the lower order tensor components, as evaluated in Eq. (5.45) for $l=4$. We observe that these terms are not obviously present for representations starting with conserved tensors (as it happens in the non-interacting theory).

8.3. Connection with the Three-Point Function

We remark that the V.E.V. selection rule (6.5) and the selection rule on three-point functions (6.25) tell us that there is a strict connection between the three-point function and the contribution of a given tensor representation to the operator product expansion (8.14) [66]. In fact, multiplying both sides of Eq. (8.14) for $O^{(n)}_{\alpha_1\ldots\alpha_n}(x'')$ and taking the vacuum expectation value, we have on the left-hand-side

$$\langle 0|O^n_{\alpha_1\ldots\alpha_n}(x'')\,A(x)\,B(x')|0\rangle \tag{8.18}$$

and on the right-hand-side, as a consequence of the V.E.V. selection rule, an integral involving the two-point function

$$\langle 0|O^n_{\alpha_1\ldots\alpha_n}(x'')\,O^n_{\beta_1\ldots\beta_n}(x')|0\rangle\,.$$

We note in particular that the selection-rule given above for an operator expansion corresponding to canonical dimensions is a consequence of the previous discussion. For example, for conserved four-tensors the three-point function (8.18) vanishes unless $l_A = l_B$.

Finally we show, as in Appendix E, how the covariant expansion given by Eq. (8.14) can be derived from the three-point function. As a consequence of the V.E.V. selection rule we limit ourselves to the scalar contribution to the expansion (8.14) and we write at $x'=0$

$$\begin{aligned} A(x)\,B(0) = (1/x^2)^{\frac{1}{2}(l_A+l_B-l)} \int_0^1 du\, u^{\frac{1}{2}(l_A-l_B+l)-1}(1-u)^{\frac{1}{2}(l_B-l_A+l)-1} \\ \cdot e^{ux\cdot\partial}\,{}_0F_1(l_n-1;\,-\tfrac{1}{4}x^2\Box\, u(1-u))\,O(0)\,. \end{aligned} \tag{8.18'}$$

On the other hand (see for instance Eq. (7.21))

$$\begin{aligned} \langle 0|C(y)\,A(x)\,B(0)|0\rangle = C_{ABC}((y-x)^{-2})^{\frac{1}{2}(l_C+l_A-l_B)} \\ \cdot (1/x^2)^{\frac{1}{2}(l_A+l_B-l_C)}(1/y^2)^{\frac{1}{2}(l_C+l_B-l_A)} \end{aligned} \tag{8.19}$$

we use the identity [8]

$$(1/y^2)^{\frac{1}{2}(l_C+l_B-l_A)}((y-x)^{-2})^{\frac{1}{2}(l_C+l_A-l_B)} = \frac{\Gamma(l_C)}{\Gamma(\frac{1}{2}(l_C+l_A-l_B))\Gamma(\frac{1}{2}(l_C+l_B-l_A))}$$

$$\cdot \int_0^1 du\, u^{\frac{1}{2}(l_C+l_A-l_B)-1}(1-u)^{\frac{1}{2}(l_C+l_B-l_A)-1}\,[(y-ux)^2]^{-l_C}$$

$$\cdot \left[1+x^2\frac{u(1-u)}{(y-ux)^2}\right]^{-l_C} = \frac{\Gamma(l_C)}{\Gamma(\frac{1}{2}(l_C+l_A-l_B))\,\Gamma(\frac{1}{2}(l_C+l_B-l_A))} \tag{8.20}$$

$$\cdot \int_0^1 du\, u^{\frac{1}{2}(l_C+l_A-l_B)-1}(1-u)^{\frac{1}{2}(l_C+l_B-l_A)-1} \sum_{h=0}^{\infty} (1/h!)\,\frac{\Gamma(l_C+h)}{\Gamma(l_C)}$$

$$\cdot(-x^2)^h\,[u(1-u)]^h\,[(y-ux)^{-2}]^{l_C+h}\,.$$

We next note that

$$\langle 0|\Box^h C(x)\,C(y)|0\rangle \propto \Box^h[(x-y)^{-2}]^{l_C}$$

$$= 4^h\,\frac{\Gamma(l_C+h)\,\Gamma(l_C-1+h)}{\Gamma(l_C)\,\Gamma(l_C-1)}\,[(x-y)^{-2}]^{l_C+h} \tag{8.21}$$

so that

$$[(y-ux)^{-2}]^{l_C+h}$$

$$= e^{ux\cdot\partial}\left(\frac{1}{4}\right)^h \frac{\Gamma(l_C)\,\Gamma(l_C-1)}{\Gamma(l_C+h)\,\Gamma(l_C-1+h)}\,\langle 0|C(y)\Box^h C(0)|0\rangle\,. \tag{8.22}$$

The contribution in Eq. (5.19) thus comes from the operator product expansion

$$\int_0^1 du\, u^{\frac{1}{2}(l_C+l_A-l_B)-1}(1-u)^{\frac{1}{2}(l_C+l_B-l_A)-1}\,e^{ux\cdot\partial}\sum_{h=0}^{\infty}(1/h!)$$

$$\cdot\frac{\Gamma(l_C-1)}{\Gamma(l_C-1+h)}(-x^2/4)^h\,(u(1-u)\Box)^h\,C(0) = \int_0^1 du\, u^{\frac{1}{2}(l_C+l_A-l_B)-1} \tag{8.23}$$

$$\cdot(1-u)^{\frac{1}{2}(l_C+l_B-l_A)-1}\,e^{ux\cdot\partial}\;{}_0F_1(l_C-1;\,-\tfrac{1}{4}x^2u(1-u)\Box)\,C(0)$$

which coincides with (8.18).

As a last step we point out some consequences of exact conformal symmetry which could be useful in future developments [67]. The arguments of this section clearly establish that the unknown coefficient C_n^{AB} which gives the contribution of the n-th-order tensor to the product of two operators (which we take for semplicity to be scalars) is directly related to the unknown coefficient C_{nAB} which normalizes the vertex function

$$\langle 0|O_{\alpha_1\ldots\alpha_n}(y)\,A(x)\,B(0)|0\rangle = C_{nAB}\mathscr{I}^{(nAB)}_{\alpha_1\ldots\alpha_n}(x,y) \tag{8.24}$$

and whose functional form is completely determined from conformal invariance. This result is a consequence of the general selection rule on two-point functions derived in Section 5. The selection rule tells us that the operator product expansion looks like an orthogonal expansion (with respect to vacuum expectation values). In fact, going for simplicity to the light-cone limit, $x^2\to 0$, we write

$$\begin{aligned} A(x)\,B(0) \underset{x^2\to 0}{\sim} & (1/x^2)^{\frac{1}{2}(l_A+l_B+n-l_n)} \sum_n C_n^{AB} x^{\alpha_1}\ldots x^{\alpha_n} \\ & \cdot \int_0^1 du\, f_n^{AB}(u)\, O^{(n)}_{\alpha_1\ldots\alpha_n}(ux) \end{aligned} \tag{8.25}$$

(the weight functions $f_n^{AB}(u)$ are given by Eq. (7.32)). Multiplying both sides for $O^{(m)}_{\beta_1\ldots\beta_m}(y)$ and taking the V.E.V. we get

$$\begin{aligned} \langle 0|O^{(n)}_{\alpha_1\ldots\alpha_n}(y)\,A(x)\,B(0)|0\rangle & \sim (1/x^2)^{\frac{1}{2}(l_A+l_B+n-l_n)}\, C_n^{AB} x^{\alpha_1}\ldots x^{\alpha_n} \\ & \cdot \int_0^1 du\, f_n^{AB}(u)\,\langle 0|O^{(n)}_{\alpha_1\ldots\alpha_n}(y)\, O^{(n)}_{\alpha_1\ldots\alpha_n}(ux)|0\rangle \end{aligned} \tag{8.26}$$

with the right-hand-side vanishing unless $m=n$ and $l_m\equiv l_n$ (orthogonality property).

If we call C_{nn} the factor normalizing the two-point function

$$\langle 0|O^{(n)}_{\beta_1\ldots\beta_n}(y)\,O^{(n)}_{\alpha_1\ldots\alpha_n}(ux)|0\rangle = C_{nn}\,G^{(nn)}_{\beta_1\ldots\beta_n\alpha_1\ldots\alpha_n}(y-ux) \tag{8.27}$$

we obtain

$$\begin{aligned} \mathscr{I}^{(nAB)}_{\beta_1\ldots\beta_n}(x,y) \underset{x^2\to 0}{=} & (1/x^2)^{\frac{1}{2}(l_A+l_B+n-l_n)}\, x^{\alpha_1}\ldots x^{\alpha_n} \\ & \cdot \int_0^1 du\, f_n^{AB}(u)\, G^{(nn)}_{\beta_1\ldots\beta_n\alpha_1\ldots\alpha_n}(y-ux)\,, \end{aligned} \tag{8.28}$$

$$C_{nAB} = C_n^{AB}\,C_{nn}\,. \tag{8.29}$$

As we have already pointed out in the introduction additional results obtain whenever gauge invariance properties apply to the theory. In general, if the tensors $O^{(n)}_{\alpha_1\ldots\alpha_n}(x)$ are conserved, or partially conserved, one can take advantage of the Ward identities associated to the vertex function in Eq. (8.24), and use them in conjunction with the relations (8.28) and (8.29), which follow from conformal invariance alone. A particularly interesting situation arises with local currents associated with symmetry groups, such as the currents j^α_μ associated to the $SU(3)\times SU(3)$ generators, or the energy momentum tensor associated to the space-time symmetry. For instance, for J_μ^α, the coefficient, C_{1AA}, in (8.24), in front of the vertex function (for $A=B$, otherwise it vanishes) is proportional to the (unitary) charge Q_A of the field $A(x)$ and to the factor C_{AA} normalizing the propagator of the A-field. Eq. (8.29) then tells us the interesting fact that the coefficient C_1^{AA}, giving in the operator product expansion for $A(x)\,A(0)$ the contribution from the local operator $J_\mu^\alpha(x)$ (together with its associated irreducible representation of the conformal algebra) is proportional to the charge Q_A and to the ratio C_{AA}/C_{11}, of the normalization factors for the AA propagator and the current-current propagator respectively, and therefore it cannot vanish unless $Q_A=0$. We have limited ourselveles only to the light-cone limit, by starting from Eq. (8.25), but clearly, with some more complication, one could as well have started from Eq. (8.16), obtaining the same results.

9. Conclusions and Outlook

In conclusion we again stress that we have carried out our discussion in two different steps: (i) validity of conformal invariance on the light cone; (ii) exact conformal invariance. In spite of its apparent weakness, assumption (i) leads to a number of important implications and connections. Among them we stress the automatic satisfaction of the causality requirement in a general conformal covariant operator product expansion on the light cone and its definite group-theoretical structure. On the other hand statements on operator expansion can be translated into statements on equal time commutators. The assumption (ii) turns out, as expected, to be very restrictive. The essential restrictions are seen in terms of the well-definite two-point function selection rule. The three-point function is fully determined in its functional form and definite selection rules again apply. The problem of the three-point function is entirely equivalent to that of the operator expansion for the product of two local operators.

In no way one has exhausted at this stage all limitations implied on the theory. Further investigation will have to analyse more closely the implications of external gauge constraints on the conformally invariant formalism, as discussed at the end of Section 8.3. Still further one needs a complete discussion of the so-called "crossing relations" resulting on the four-point function when its four local operators are in various ways associated in pairs, Wilson-expanded, and the different outcomes compared. But already at the initial stages of the investigation, as described in the present review, it appears that conformal invariance, in that limit in which it is thought to possibly apply to physical theories, is capable of strongly reducing the dynamical complexity.

Appendices

Appendix A. Restrictions imposed by conformal symmetry on the equal time commutator of two local operators.

Using Eq. (4.1) and the Jacobi identity with D, together with Eq. (4.6), we find

$$\begin{aligned}\delta(nx)\,[[A_r(x)\,B_s(0)],\,D] &= \sum_k i^k [S^{\tau_1\ldots\tau_k}(A_r,B_s,0),D]\,\partial_{\tau_1}\ldots\partial_{\tau_k}\delta(x)\\ &= \delta(nx)\,[A_r(x),[B_s(0),D]]+[[A_r(x),D],B_s(0)]\\ &= -\sum_k i^{k+1}[(l_A+l_B)\,S^{\tau_1\ldots\tau_k}(A_r,B_s,0)\\ &\quad - x^\varrho S^{\tau_1\ldots\tau_k}(\partial_\varrho A_r,B_s,0)]\,\partial_{\tau_1}\ldots\partial_{\tau_k}\delta(x)\\ &= -\sum_k i^{k+1}[(l_A+l_B)\,S^{\tau_1\ldots\tau_k}(A_r,B_s,0)\\ &\quad + i(k+1)\,S^{\lambda\tau_1\ldots\tau_k}(\partial_\lambda A_r,B_s,0)]\,\partial_{\tau_1}\ldots\partial_{\tau_k}\delta(x)\,.\end{aligned}$$

The result one obtains is therefore

$$\begin{aligned}&i[S^{\tau_1\ldots\tau_k}(A_r,B_s,n,x),D]\\ &\quad =(l_A+l_B)\,S^{\tau_1\ldots\tau_k}(A_r,B_s,n,x)+i(k+1)\,S^{\lambda\tau_1\ldots\tau_k}(\partial_\lambda A_r,B_s,n,x)\,.\end{aligned}\tag{A.1}$$

Eqs. (4.1), (4.7) and the Jacobi identity with K_μ give us

$$\begin{aligned}\delta(nx)\,[[A_r(x),B_s(0)],K_\mu] &= \sum_k i^k [S^{\tau_1\ldots\tau_k}(A_r,B_s,0),K_\mu]\\ &\quad\cdot\partial_{\tau_1}\ldots\partial_{\tau_k}\delta(x)=\delta(nx)\{[A_r(x),[B_s(0),K_\mu]]+[[A_r(x),K_\mu],B_s(0)]\}\\ &= \sum_k i^{k+1}[-2l_A x_\mu S^{\tau_1\ldots\tau_k}(A_r,B_s,0)+(2x_\mu x^\varrho - x^2 g^\varrho_\mu)\\ &\quad\cdot S^{\tau_1\ldots\tau_k}(\partial_\varrho A_r,B_s,0)-2ix^\varrho S^{\tau_1\ldots\tau_k}((\Sigma_{\mu\varrho}A)_r,B_s,0)]\,\partial_{\tau_1}\ldots\partial_{\tau_k}\delta(x)\\ &= \sum_k i^{k+1}[2il_A(k+1)\,g_{\mu\lambda}S^{\lambda\tau_1\ldots\tau_k}(A_r,B_s0)-(k+2)\,(k+1)\\ &\quad\cdot(2g_{\mu\lambda_1}g^\varrho_{\lambda_2}\cdot g_{\lambda_1\lambda_2}g^\varrho_\mu)\,S^{\lambda_1\lambda_2\tau_1\ldots\tau_k}(\partial_\varrho A_r,B_s0)\\ &\quad + 2(k+1)\,S^{\lambda\tau_1\ldots\tau_k}((\Sigma_{\mu\nu}A)_r,B_s0)]\,.\end{aligned}$$

One can thus write

$$\begin{aligned}&-\tfrac{1}{2}i(k+2)\,(2g^\mu_{\lambda_1}g^\varrho_{\lambda_2}-g_{\lambda_1\lambda_2}g^{\mu\varrho})\,S^{\lambda_1\lambda_2\tau_1\ldots\tau_k}(\partial_\varrho A_r,B_s,n,x)\\ &\quad = l_A S^{\mu\tau_1\ldots\tau_k}(A_r,B_s,n,x)+iS^{\lambda\tau_1\ldots\tau_k}((\Sigma^{\mu\nu}A)_r,B_s,n,x)\\ &\qquad+\tfrac{1}{2}(k+1)^{-1}[S^{\tau_1\ldots\tau_k}(A_r,B_s,n,x),K^\mu]\,.\end{aligned}\tag{A.2}$$

We shall now look for simpler expressions in place of Eqs. (A.1) and (A.2). We take Eq. (4.24) for the value $k+1$ and multiply by $g_{\nu,k+1}$. We obtain:

$$iS^{\lambda\tau_1\dots\tau_k}(\partial_\lambda A_r, B_s, 0) = (k+1)^{-2}\sum_{J=1}^{k} kS^{\tau_1\dots\tau_k}(A_r, B_s, 0)$$
$$+ \mathrm{d}/\mathrm{d}n_\lambda)n_\lambda S^{\tau_1\dots\tau_k}(A_r, B_s, 0) + 4kS^{\tau_1\dots\tau_k}(A_r, B_s, 0)$$
$$= \frac{k+3}{k+1} S^{\tau_1\dots\tau_k}(A_r, B_s, 0)\,.$$

Using now Eq. (4.31) together with the above equation one derives

$$(l_A + l_B + k + 3)\, S^{\tau_1\dots\tau_k}(A_r, B_s, 0) = i[S^{\tau_1\dots\tau_k}(A_r, B_s, 0), D]\,. \tag{A.3}$$

The result states nothing else than the obvious consequence that $S^{\tau_1\dots\tau_k}$ must transform covariantly with respect to D and it has scale dimensions $l_A + l_B + k + 3$. Since all operators have negative scale dimensions (in a scale invariant theory) it follows that the highest order of the Schwinger terms in a commutator of fields of definite scale dimensions is fixed. Now, to simplify Eq. (A.2), we write its left-hand side as

$$-i\tfrac{1}{2}(k+2)\,[S^{\mu\varrho\tau_1\dots\tau_k}(\partial_\varrho A_r, B_s, 0) - g_{\lambda_1\lambda_2} S^{\lambda_1\lambda_2\tau_1\dots\tau_k}(\partial^\mu A_r, B_s, 0)]$$
$$+ i\tfrac{1}{2}(k+2)\, S^{\mu\varrho\tau_1\dots\tau_k}(\partial_\varrho A_r, B_s, 0)\,.$$

We than use Eq. (4.23) to calculate the term within square brackets in the above equation, which becomes

$$-\tfrac{1}{2}i\{[S^{\lambda\tau_1\dots\tau_k}(A_r, B_s, 0), M_{\mu\lambda}] - S^{\lambda\tau_1\dots\tau_k}((\Sigma_{\mu\lambda}A)_r, B_s, 0)$$
$$- S^{\lambda\tau_1\dots\tau_k}(A_r, (\Sigma_{\mu\lambda}B)_s, 0)\}\,. \tag{A.4}$$

On the other hand, using Eq. (4.24), one has

$$-i\tfrac{1}{2}(k+2)\, S^{\mu\varrho\tau_1\dots\tau_k}(\partial_\mu A_r, B_s, 0) = -\tfrac{1}{2}(k+4)\, S^{\mu\tau_1\dots\tau_k}(A_r, B_s, 0) \tag{A.5}$$

Eqs. (A.2), (A.4) and (A.5) give

$$(k+1)^{-1}[S^{\tau_1\dots\tau_k}(A_r, B_s, 0), K^\mu] = (2l_A + k + 4)\, S^{\mu\tau_1\dots\tau_k}(A_r, B_s, 0)$$
$$+ iS^{\lambda\tau_1\dots\tau_k}((\Sigma^\mu_\lambda A)_r, B_s, 0) - iS^{\lambda\tau_1\dots\tau_k}(A_r, (\Sigma^\mu_\lambda B)_s, 0) \tag{A.6}$$
$$+ i[S^{\lambda\tau_1\dots\tau_k}(A_r, B_s, 0), M^\mu_\lambda]\,.$$

The left-hand-side of Eq. (A.6) vanishes provided $S^{\tau_1\dots\tau_k}(A_r, B_s, 0)$ is a superposition of local operators transforming accoding to finite dimensional representations with $K_\mu = 0$ of the conformal group. Also, one gets from Eq. (A.6) the result $[S^{\tau_1\dots\tau_N}(A_r, B_s, 0), K_\mu] = 0$ (N being the order of the last Schwinger term in Eq. (4.1)). From Eq. (4.13) one has

$$-(k+1)^{-1}[S^{\tau_1\dots\tau_k}(A_r, B_s, 0) + (-1)^k S^{\tau_1\dots k}(B_s, A_r, 0), K^\varrho]$$
$$= (k+1)^{-1}\sum_{J=k+1} i^{J-k}\binom{J}{k}(\partial_{\tau_k}\dots\partial_{\tau_{J-k}}[S^{\tau_1\dots\tau_k}(A_r, B_s, x), K^\mu])_{x=0} \tag{A.7}$$

whose left-hand-side can be written, using Eqs. (A.6) and (4.13) in the form

$$\begin{aligned}&\{[k+4+l_A+l_B]\,g^{\mu}_{\lambda}+\mathrm{ad}\,M^{\mu}_{\lambda}\}\,S[S^{\lambda\tau_1\ldots\tau_k}(A_r,B_s,0)-(-1)^k S^{\lambda\tau_1\ldots\tau_k}(B_s,A_r,0)]\\&\quad-[(l_A-l_B)\,\delta^{r'}_r\,\delta^{s'}_s+i(\Sigma^{\mu}_{\lambda})^{r'}_r\,\delta^{s'}_s-i(\Sigma^{\mu}_{\lambda})^{s'}_s\,\delta^{r'}_r]\cdot\\&\quad\cdot\sum_{J=k+2}\binom{J}{k+1}\,i^{J-k+1}\,\partial_{\tau_{k+1}}\ldots\partial_{\tau_J}S^{\lambda\tau_1\ldots\tau_J}(A_r,B_s,0)\,.\end{aligned}\tag{A.8}$$

The notation is usual in Lie algebras: $\mathrm{ad}\,M_{\mu\nu}S$ stands for $[S,M_{\mu\nu}]$. When we compare Eq. (A.8) to Eq. (A.7) we find

$$\begin{aligned}&i\partial_{\lambda}[S^{\lambda\tau_1\ldots\tau_{N-1}}(A_r,B_s,0),K^{\mu}]\\&\quad=\{(N+3+l_A+l_B)\,g^{\mu}_{\lambda}+i\,\mathrm{ad}\,M^{\mu}_{\lambda}\}\,2S^{\lambda\tau_1\ldots\tau_{N-1}}(A_r,B_s,0)\end{aligned}\tag{A.9}$$

In deriving Eq. (A.9) we have used the equation

$$S^{\tau_1\ldots\tau_N}(A_r,B_s,0)=(-1)^N S^{\tau_1\ldots\tau_N}(B_s,A_r,0)\,.$$

Eq. (A.9) can be written, recalling Eq. (A.3), as

$$\begin{aligned}&-i\partial_{\lambda}[S^{\lambda\tau_1\ldots\tau_{N-1}}(A_r,B_s,0),K^{\mu}]\\&\quad=2i\{g^{\mu}_{\lambda}\,\mathrm{ad}\,D+\mathrm{ad}\,M^{\mu}_{\lambda}\}\,S^{\lambda\tau_1\ldots\tau_{N-1}}(A_r,B_s,0)\end{aligned}\tag{A.10}$$

which is satisfied identically. The only independent restriction one obtains, for $k=N-1$, from the result in Eq. (A.6) is

$$\begin{aligned}&-N^{-1}[S^{\tau_1\ldots\tau_{N-1}}(A_r,B_s,0)-(-1)^{N-1}S^{\tau_1\ldots\tau_{N-1}}(B_s,A_r,0),K_{\mu}]\\&\quad=2[(l_A-l_B)\,g_{\mu\lambda}\delta^{r'}_r\,\delta^{s'}_s+i(\Sigma_{\mu\lambda})^{r'}_r\,\delta^{s'}_s-i(\Sigma_{\mu\lambda})^{s'}_s\,\delta^{r'}_r]\\&\quad\cdot S^{\lambda\tau_1\ldots\tau_{N-1}}(A_r,B_s,0)\,.\end{aligned}\tag{A.11}$$

It is interesting to note that Eq. (A.6) is identically satisfied if one takes for A the operator $n^{\varrho}\theta_{\varrho\sigma}$ and if in the commutator $[\theta_{\mu\nu},B]$ the value of N is $=1$.

Appendix B. Properties of the tensors $\psi_{A_1\ldots A_n}$.

Let us consider an operator $\psi_{A_1\ldots A_n}(\eta)$, transforming as a tensor under $O(4,2)$ transformations, defined on the cone $\eta^2=0$, of degree of homogeneity $\lambda_n=-l_n$, i.e. satisfying

$$\eta^A\partial_A\psi_{A_1\ldots A_n}(\eta)=\lambda_n\psi_{A_1\ldots A_n}(\eta)\,.\tag{B.1}$$

It is easy to show that the orbital quadratic Casimir reduces on the cone to

$$L_{AB}L^{AB}=2\eta^A\partial_A(4+\eta^A\partial_A)\tag{B.2}$$

(the other two Casimir operators vanish for orbital representations) and therefore

$$L_{AB}L^{AB}\psi_{A_1\ldots A_n}(\eta)=2l_n(l_n-4)\,\psi_{A_1\ldots A_n}(\eta)\,.\tag{B.3}$$

The equivalence between (5.15a) and (5.16a) is thus proved. The statements (5.15b) and (5.16b) are obviously equivalent.

In order to prove the equivalence between (5.15c) and (5.16c) we observe that

$$J^{AB}J_{AB}=L^{AB}L_{AB}+S^{AB}S_{AB}+2L^{AB}S_{AB} \tag{B.4}$$

where

$$(S_{AB}\psi)_{A_1\dots A_n}(\eta)=i\sum_{i=1}^{n}\left(g_{AA_i}\psi_{A_1\dots\hat{A}_i\dots A_nB}(\eta)-g_{BA_i}\psi_{A_1\dots\hat{A}_i\dots A_nA}(\eta)\right). \tag{B.5}$$

Making use of the two supplementary conditions (5.15c) we have

$$\begin{aligned}(L-S\psi)_{A_1\dots A_n}(\eta)&\\ &=-2\sum_{i=1}^{n}\left(\eta_{A_i}\partial^B\psi_{A_1\dots\hat{A}_i\dots A_nB}(\eta)-\eta^A\partial^{A_i}\psi_{A_1\dots\hat{A}_i\dots A_nA}(\eta)\right)\\ &=2\sum_{i=1}^{n}\eta^A\partial^{A_i}\psi_{A_1\dots\hat{A}_i\dots A_n}(\eta)=-2n\psi_{A_1\dots A_n}(\eta).\end{aligned} \tag{B.6}$$

Moreover we have[8] [67]

$$\begin{aligned}&\pi\cdot k=n\\ &S_{AB}S^{AB}=2n(n+4)\\ &\varepsilon_{ABCDEF}S^{AB}S^{CD}S^{EF}=0\\ &S_A^B S_B^C S_C^D S_D^A=3n(n+2).\end{aligned} \tag{B.7}$$

Appendix C. Proof of Lemma 2 and construction of the mapping of $\psi_{A_1\dots A_n}$ into irreducible representations of conformal algebra.

Let us prove Lemma 2: remembering [see (5.12)] that

$$O_{\alpha_1\dots\alpha_n}(x)=k^{l_n}(e^{-ix\cdot\pi})^{A_1\dots A_n}_{\alpha_1\dots\alpha_n}\psi_{A_1\dots A_n}(\eta) \tag{C.1}$$

and noting that

$$\tfrac{1}{2}L^{AB}L_{AB}=l_n(l_n-4), \tag{C.2}$$

$$\tfrac{1}{2}S_{AB}S^{AB}=\tfrac{1}{2}S_{\mu\nu}S^{\mu\nu}+\pi\cdot k-S_{65}^2+4iS_{65}, \tag{C.3}$$

$$S_{AB}L^{AB}=S_{\mu\nu}L^{\mu\nu}-2S_{65}L_{65}+\pi\cdot\mathscr{K}+k\cdot\mathscr{P}, \tag{C.4}$$

where $\mathscr{K}_\mu=L_{6\mu}-L_{5\mu}$, $\mathscr{P}_\mu=L_{6\mu}+L_{5\mu}$ we get, in terms of the generators transformed by $e^{-ix\pi}\dots e^{ix\pi}$,

$$\begin{aligned}e^{-ix\pi}J_{AB}J^{AB}e^{ix\pi}=&\,l_n(l_n-4)+\tfrac{1}{2}S_{\mu\nu}S^{\mu\nu}+\pi\cdot\tilde{k}+\tilde{S}_{65}^2+4i\tilde{S}_{65}\\ &+\tilde{S}_{\mu\nu}\tilde{L}^{\mu\nu}-2\tilde{S}_{65}\tilde{L}_{65}+\pi\cdot\tilde{\mathscr{K}}+\tilde{k}\cdot\tilde{\mathscr{P}},\end{aligned} \tag{C.5}$$

---

[8] Note that S_{65} has vanishing eigenvalue on the highest order Lorentz tensor contained in the representation.

$$\tilde{S}_{\mu\nu} = S_{\mu\nu} + x_\mu \pi_\nu - x_\nu \pi_\mu, \qquad \tilde{S}_{65} = S_{65} + x \cdot \pi$$

$$\tilde{K}_\mu = k_\mu + 2x^\nu (g_{\mu\nu} S_{65} + S_{\mu\nu}) + 2x_\mu x \cdot \pi - x^2 \pi_\mu \tag{C.6}$$

$$\tilde{L}_{\mu\nu} = L_{\mu\nu} + x_\nu \pi_\mu - x_\mu \pi_\nu, \qquad \tilde{\mathscr{P}}_\lambda = \mathscr{P}_\lambda - \pi_\lambda$$

$$\tilde{\mathscr{K}}_\lambda = \mathscr{K}_\lambda + 2il_n x_\lambda - 2x_\lambda x \cdot \pi + x^2 \pi_\lambda$$

$$\tilde{L}_{65} = il_n - x \cdot \pi$$

$$\tilde{S}^2_{65} = S^2_{65} + 2x \cdot \pi + (x \cdot \pi)^2 .$$

Using translation invariance it follows that

$$\begin{aligned} \tfrac{1}{2} J^{AB} J_{AB} O_{\alpha_1 \ldots \alpha_n}(0) &= [l_n(l_n - 4) + \tfrac{1}{2} S_{\mu\nu} S^{\mu\nu} + \pi \cdot k + k \cdot (\mathscr{P} - \pi)] \, O_{\alpha_1 \ldots \alpha_n}(0) \\ &= [l_n(l_n - 4) + n(n+2)] \, O_{\alpha_1 \ldots \alpha_n}(0) \end{aligned} \tag{C.7}$$

where we used the properties

$$(S_{65} O)_{\alpha_1 \ldots \alpha_n}(0) = 0 , \tag{C.8}$$

$$(k_\lambda O)_{\alpha_1 \ldots \alpha_n}(0) = 0 . \tag{C.9}$$

Eq. (C.9) is a consequence of the first supplementary condition (5.15c).

In order to complete the proof of Lemma 2 the mapping between covariant tensors $\psi_{A_1 \ldots A_n}(\eta)$ and irreducible representations of conformal algebra on space-time of the type discussed in Chapter 2 and Lemma 1 must be exhibited.

We show that, if $l_n \neq 2 + n$, the components of the tensor

$$\psi_{A_1 \ldots A_n}(\eta) = k^{-l_n} (e^{ix \cdot \pi} O)_{A_1 \ldots A_n}(x) \tag{C.10}$$

are completely specified in terms of the divergences of the highest order tensor $O_{\alpha_1 \ldots \alpha_n}(x)$ [66, 67]. As $e^{ix \cdot \pi} = 1$ for $x = 0$, it is sufficient to evaluate the components

$O_{\alpha_1 \ldots \alpha_J xx \ldots}(0)$ where $(J = 0, 1, \ldots, n)$ and $xx \ldots$ stands for 5 or 6 .

This problem can be completely solved by using the two supplementary conditions (5.15c).

To obtain the components of $O_{A_1 \ldots A_n}(0)$ we first observe that:

$$(S_{65} O)_{A_1 \ldots A_n} = i \sum_{i=1}^{n} (g_{5A_i} O_{A_1 \ldots \hat{A}_i \ldots A_n 6} - g_{6A_i} O_{A_1 \ldots \hat{A}_i \ldots A_n 5}) \tag{C.11}$$

and using the first supplementary condition at $x = 0$ (we omit the index 0) we get

$$O_{5 A_2 \ldots A_n} = O_{6 A_2 \ldots A_n} . \tag{C.12}$$

Eq. (C.11) becomes

$$(S_{65} O)_{A_1 \ldots A_n} = i \sum_{i=1}^{n} (g_{5A_1} - g_{6A_i}) \, O_{A_1 \ldots \hat{A}_i \ldots A_n 6} \tag{C.13}$$

so

$$(S_{65}O)_{\alpha_1\ldots\alpha_k A_{k+1}\ldots A_n} = i\sum_{k+1}^{n}(g_{5A_i}-g_{6A_i})O_{\alpha_1\ldots\alpha_k A_{k+1}\ldots\hat{A}_i\ldots A_n 6}$$
$$= -i\sum_{k+1}^{n}(\delta_{5A_i}+\delta_{6A_i})\,O_{\alpha_1\ldots\alpha_k A_{k+1}\ldots A_i\ldots A_n 6}\,. \tag{C.14}$$

In $A_{k+1}, \ldots, A_n$ indices 5 can be changed into 6 or viceversa because of condition (C.12). We get:

$$(S_{65}O)_{\alpha_1\ldots\alpha_n 6\ldots 6} = -i(n-k)\,O_{\alpha_1\ldots\alpha_k 6\ldots 6} \tag{C.15}$$

and, defining $L = iS_{65}$,

$$(LO)_{\alpha_1\ldots\alpha_k 6\ldots 6} = (n-k)\,O_{\alpha_1\ldots\alpha_k 6\ldots 6}\,. \tag{C.16}$$

So all Lorentz tensor are eigenstates of the dimension.

Remembering Eq. (B.5) we have:

$$(\pi_\mu O)_{A_1\ldots A_n} = i\sum_{i=1}^{n}[(g_{6A_i}+g_{5A_i})\,O_{A_1\ldots\hat{A}_i\ldots A_n\mu}$$
$$-g_{\mu A_i}(O_{A_1\ldots\hat{A}_i\ldots A_n 6}+O_{A_1\ldots\hat{A}_i\ldots A_n 5})] \tag{C.17}$$

$$(i\pi_\mu O)^\mu_{A_2\ldots A_n} = 2(4O_{6A_2\ldots A_n}+\sum_{i=2}^{n}g_{\mu A_i}O^\mu_{A_2\ldots\hat{A}_i\ldots A_n 6})$$
$$= 2\left[(4+n-1)\,O_{A_2\ldots A_n 6}+\sum_{i=2}^{n}(g_{5A_i}O_{5A_2\ldots\hat{A}_i\ldots A_n}-g_{6A_i}O_{6A_2\ldots\hat{A}_i\ldots A_n})\right] \tag{C.18}$$

$$(i\pi_\mu O)^\mu_{\alpha_2\ldots\alpha_k A_{k+1}\ldots A_n} = 2(4+n-1)O_{\alpha_2\ldots\alpha_k A_{k+1}\ldots A_n 6}$$
$$-\sum_{i=k+1}^{n}(\delta_{5A_i}+\delta_{6A_i})\,O_{6\alpha_2\ldots\alpha_k A_{k+1}\ldots A_i\ldots A_n} \tag{C.19}$$
$$= 2(3+k)\,O_{\alpha_2\ldots\alpha_k 6\ldots 6}(A_{k+1}\,\ldots\,A_n = 6\,\ldots\,6)\,.$$

Using (C.19) and the second supplementary condition at $x=0$ we get[9] (we recall that $\partial/\partial\eta^A = k^{-1}(g^\mu_A-[g^6_A-g^5_A]\,x^\mu)\,\partial/\partial x_\mu+(g^6_A-g^5_A)\,\partial/\partial k + 2\eta_A\partial/\partial\eta^2$)

$$\tfrac{1}{2}\partial_\mu O^\mu_{\alpha_2\ldots\alpha_k 6\ldots 6}+(2+k-l_n)\,O_{\alpha_2\ldots\alpha_k 6\ldots 6}=0\,, \qquad 2\leqq k\leqq n\,. \tag{C.20}$$

Eq. (C.20) can be easily solved by iteration and we obtain

$$O_{\alpha_1\ldots\alpha_{n-k}6\ldots 6}(x) = 2^{-k}\,\frac{\Gamma(l_n-2-n)}{(l_n-2-n+k)}\,\partial_{\mu_1}\ldots\partial_{\mu_k}O^{\mu_1\ldots\mu_k}_{\alpha_1\ldots\alpha_{n-k}}(x) \tag{C.21}$$

which solves the problem and completes the proof of Lemma 2.

---

[9] The generalized Lorentz condition $\partial^A\,\psi_{AA_2\ldots A_n}(\eta)=0$ can be rewritten as $(L^A_B - ig^A_B)\cdot\psi_{AA_2\ldots A_n}(\eta)=0$ which is manifestly defined on the cone $\eta^2=0$.

We observe that for correlated dimensions $l_n = l + n$ formula (C.21) simplifies to

$$O_{\alpha_1 \ldots \alpha_{n-k} 6 \ldots 6}(x) = 2^{-k} \frac{\Gamma(l-2)}{(l-2+k)} \partial_{\mu_1} \ldots \partial_{\mu_k} O^{\mu_1 \ldots \mu_k}_{\alpha_1 \ldots \alpha_{n-k}}(x). \tag{C.22}$$

Appendix D. Operator product decomposition in a continuous basis.

It is particularly interesting to deduce the analogue of Eq. (7.22) on the continuous basis. In this case, instead of the expansion (7.17) we have

$$A(x)\, B(0) \underset{x^2 \to 0}{=} \sum_{n=0}^{\infty} (1/x^2)^{\frac{1}{2}(l_A + l_B + n - l_n)} x^{\alpha_1} \ldots x^{\alpha_n} \int_0^1 du\, f_n^{AB}(u)\, O_{\alpha_1 \ldots \alpha_n}(ux). \tag{D.1}$$

With the same procedure as in the discrete basis, commuting with K_λ we have

$$\begin{aligned}
(-i)\, [A(x)\, B(0), K_\lambda] &= (2x_\lambda x \cdot \partial - x^2 \partial_\lambda + 2 l_A x_\lambda)\, A(x)\, B(0) \\
&= \sum_{n=0}^{\infty} (1/x^2)^{\frac{1}{2}(l_A + l_B + n - l_n)} [(l_A - l_B + l_n + n)]\, x_\lambda x^{\alpha_1} \ldots x^{\alpha_n} \\
&\quad - x^2 \sum_{i=1}^{n} g^{\alpha_i}_{\lambda} x^{\alpha_i} \ldots x^{\hat{\alpha}_i} \ldots x^{\alpha_n}) \int_0^1 du\, f_n^{AB}(u)\, O_{\alpha_1 \ldots \alpha_n}(ux) \\
&\quad + x^{\alpha_1} \ldots x^{\alpha_n} \int_0^1 du\, f_n^{AB}(u)\, (2x_\lambda x \cdot \partial - x^2 \partial_\lambda)\, O_{\alpha_1 \ldots \alpha_n}(ux)
\end{aligned} \tag{D.2}$$

[after insertion of the r.h.s. of Eq. (D.1)]

$$\begin{aligned}
&= \sum_{n=0}^{\infty} (1/x^2)^{\frac{1}{2}(l_A + l_B + n - l_n)} C_n^{AB} x^{\alpha_1} \ldots x^{\alpha_n} \int_0^1 du\, f_n^{AB}(u) \\
&\quad \cdot (2x_\lambda x \cdot \partial - x^2 \partial_\lambda + 2 l_n x_\lambda - 2i x^\nu \Sigma_{\lambda\nu})\, O_{\alpha_1 \ldots \alpha_n}(ux) \qquad (\partial_\lambda = \partial/\partial x_\lambda) \\
&= \sum_{n=0}^{\infty} (1/x^2)^{\frac{1}{2}(l_A + l_B + n - l_n)} \Bigg[x^{\alpha_1} \ldots x^{\alpha_n} \int_0^1 du\, f_n^{AB}(u) \\
&\quad \cdot u(2x_\lambda x \cdot \partial - x^2 \partial_\lambda + 2(l_n + n)\, x_\lambda)\, O_{\alpha_1 \ldots \alpha_n}(ux) \\
&\quad - 2x^2 \sum_{i=1}^{n} g^{\alpha_i}_{\lambda} x^{\alpha_1} \ldots x^{\hat{\alpha}_i} \ldots x^{\alpha_n} \int_0^1 du\, u f_n^{AB}(u)\, O_{\alpha_1 \ldots \alpha_n}(ux) \Bigg]
\end{aligned}$$

where we have used the property

$$\begin{aligned}
-2i x^{\alpha_1} \ldots x^{\alpha_n} x^\nu (\Sigma_{\lambda\nu} O_{\alpha_1 \ldots \alpha_n}(ux) &= 2n x_\lambda x^{\alpha_1} \ldots x^{\alpha_n} O_{\alpha_1 \ldots \alpha_n}(ux)) \\
&\quad - 2x^2 \sum_{i=1}^{n} g^{\alpha_i}_{\lambda} x^{\alpha_1} \ldots x^{\hat{\alpha}_i} \ldots x^{\alpha_n} O_{\alpha_1 \ldots \alpha_n}(ux)
\end{aligned} \tag{D.3}$$

Comparing the most dominant terms for $x^2 \to 0$ in Eq. (D.2) we obtain

$$\begin{aligned}
(l_A - l_B + l_n + n) \int_0^1 du\, f_n^{AB}(u)\, O_{\alpha_1 \ldots \alpha_n}(ux) + 2 \int_0^1 du\, f_n^{AB}(u)\, (x \cdot \partial)\, O_{\alpha_1 \ldots \alpha_n}(ux) \\
= 2(l_n + n) \int_0^1 du\, u f_n^{AB}(u)\, O_{\alpha_1 \ldots \alpha_n}(ux) + 2 \int_0^1 du\, u f_n^{AB}(u)\, (x \cdot \partial)\, O_{\alpha_1 \ldots \alpha_n}(ux)
\end{aligned} \tag{D.4}$$

and $x\partial = u(\mathrm{d}/\mathrm{d}u)$. Integration by parts yields

$$
\begin{aligned}
(l_A - l_B + l_n + n) \int_0^1 \mathrm{d}u f_n^{AB}(u)\, O_{\alpha_1\ldots\alpha_n}(ux) - 2 \int_0^1 \mathrm{d}u (\mathrm{d}/\mathrm{d}u)\, (u f_n^{AB}(u)) \\
\cdot O_{\alpha_1\ldots\alpha_n}(ux) = 2(l_n + n) \int_0^1 \mathrm{d}u f_n^{AB}(u)\, u O_{\alpha_1\ldots\alpha_n}(ux) \\
- 2 \int_0^1 \mathrm{d}u (\mathrm{d}/\mathrm{d}u)\, (u^2 f_n^{AB}(u))\, O_{\alpha_1\ldots\alpha_n}(ux)
\end{aligned}
\tag{D.5}
$$

and comparing the integrands of both sides we finally obtain the differential equation

$$
u(1-u)\,(\mathrm{d}/\mathrm{d}u) f_n^{AB}(u) = [\tfrac{1}{2}(l_A - l_B + l_n + n) - 1 + u(2 - l_n - n)]\, f_n^{AB}(u) \tag{D.6}
$$

which is exactly the analogous of Eq. (7.22) on the continuous basis. Its general solution is given exactly by Eq. (7.32) apart from a multiplicative constant. We also remark that the causality condition turns out to be a support property of the Clebsh-Gordon coefficients $f_n^{AB}(u)$ [64].

Appendix E. Conformal covariant light-cone expansion and three point functions.

We want to point out the relation between the conformally covariant light-cone expansion derived in Chapter 7 and the results on three-point functions described in Chapter 6 [128, 65].

Using the conformally invariant solution for the three-point function as given by Eq. (6.21) we obtain, in the light-cone limit $x^2 \to 0$

$$
\begin{aligned}
\langle 0|C(y)\, A(x)\, B(0)|0\rangle \underset{x2 \to 0}{=} C_{ABC}(1/x^2)^{\frac{1}{2}(l_A + l_B - l_C)} \\
\cdot (1/y^2)^{\frac{1}{2}(l_B + l_C - l_A)}\, [(y^2 - 2xy)^{-1}]^{\frac{1}{2}(l_A + l_C - l_B)}\,.
\end{aligned}
\tag{E.1}
$$

On the other hand we can express the product $A(x)\,B(0)$, for $x^2 \to 0$ using Eq. (7.30), and obtain for the left-hand side of Eq. (E.1)

$$
\begin{aligned}
\text{L.H.S.} = (1/x^2)^{\frac{1}{2}(l_A + l_B - l_C)} \int_0^1 \mathrm{d}u\, u^{\frac{1}{2}(l_A - l_B + l_C) - 1} \\
\cdot (1-u)^{\frac{1}{2}(l_B - l_A + l_C) - 1} \langle 0|C(y)\, C(ux)|0\rangle
\end{aligned}
\tag{E.2}
$$

where we have used the selection-rules on two-point-functions derived in Chapter 6 and (Eq. (6.4))

$$
\langle 0|C(y)\, C(ux)|0\rangle \propto [(y - ux)^{-2}]^{l_C} \simeq [(y^2 - 2ux - y)^{-1}]^{l_C}\,. \tag{E.3}
$$

Inserting Eq. (E.3) in Eq. (E.2), we note that the integral appearing in (E.2) is just the Riemann-Liouville integral representation of the quantity [8, 128]

$$
(1/y^2)^{\frac{1}{2}(l_C + l_B - l_A)}\, [(y^2 - 2x - y)^{-1}]^{\frac{1}{2}(l_C + l_A - l_B)}\,. \tag{E.4}
$$

We also remark that one can calculate the contribution of a given tensor representation of the conformal algebra, contained in the product

$A(x)\,B(0)$, to the off-mass-shell vertex function. One has:

$$V_n(x^2, x\cdot p) \underset{x^2\to 0}{=} \langle 0|A(x)\,A(0)|p\rangle = (\text{constant})\,(1/x^2)^{\frac{1}{2}(l_A+l_B+n-l_n)}(x\cdot p)^n \cdot {}_1F_1\left(\tfrac{1}{2}(l_A-l_B+l_n+n); l_n+n; -ix\cdot p\right). \tag{E.5}$$

In Eq. (E.5) $|p\rangle$ is a scalar state (or spin averaged) and the subscript n applies to the given representation.

References

1. Albrecht, W., et al.: DESY Preprint 69146 (1969).
2. Bardacki, K., Segrè, G.: Phys. Rev. **159**, 1263 (1967).
3. Bateman Manuscript Project: Vol. I, Chapter VI. New York: McGraw-Hill 1900.
4. Bateman Manuscript Project: Vol. I, Chapter VI, 19.255. New York: McGraw-Hill 1900.
5. Bateman Manuscript Project: Vol. I, Chapter VI, 19.253. New York: McGraw-Hill 1900.
6. Bateman Manuscript Project: Vol. I, Chapter VI, 19.265. New York: McGraw-Hill 1900.
7. Bateman Manuscript Project: Vol. I, Chapter VI, 19.268. New York: McGraw-Hill 1900.
8. Bateman Manuscript Project: Integral transforms, Vol. II, 19.186. New York: McGraw-Hill 1900.
9. Belinfante, F. J.: Physica **6**, 887 (1939).
10. Belinfante, F. J.: Physica **7**, 305 (1940).
11. Bjorken, J. D.: Phys. Rev. **148**, 1467 (1966).
12. Bjorken, J. D.: In: Steinberger, J., (Ed.): Proc. Intern. School of Physics „Enrico Fermi". Course 41. New York: Academic Press 1969.
13. Bjorken, J. D.: Phys. Rev. **179**, 1547 (1969).
14. Bjorken, J. D., Paschos, E. A.: Phys. Rev. **185**, 1975 (1969).
15. Bloom, E. D., et al.: Phys. Rev. Letters **23**, 930 (1969).
16. Bohm, D., Flato, M., Sternheimer, D., Vigier, J. P.: Nuovo Cimen to **38**, 1941 (1965).
17. Boulware, D. G., Deser, S.: J. Math. Phys. **8**, 1468 (1967).
18. Boulware, D. G., Brown, L. S., Peccei, R. D.: Phys. Rev. D**2**, 293 (1970).
19. Boulware, D. G., Brown, L. S., Peccei, R. D.: University of Washing Preprint RLO-1388-598 (1971).
20. Brandt, R. A.: Ann. Phys. **44**, 221 (1967).
21. Brandt, R. A.: Phys. Rev. Letters **22**, 1795 (1968).
22. Brandt, R. A.: Phys. Rev. Letters **23**, 1260 (1969).
23. Brandt, R. A.: Phys. Rev. D**1**, 2808 (1970).
24. Brandt, R. A., Preparata, G.: Phys. Rev. D**1**, 2577 (1970).
25. Brandt, R. A., Preparata, G.: Nucl. Phys. B**27**, 541 (1971).
26. Breidenback, M., et al.: Phys. Rev. Letters **23**, 935 (1969).
27. Brown, L. S.: Phsy. Rev. **150**, 1338 (1966).
28. Brown, L. S.: Phys. Rev. **158**, 1444 (1967).
29. Brown, L. S.: Lectures given at the Summer Institute for Theoretical Physics, University of Colorado, Boulder (1969), to be published.
30. Budagov, I., et al.: Phys. Letters **30**B, 364 (1969).
31. Callan Jr., C. G.: Phys. Rev. D**2**, 1541 (1970).
32. Callan Jr., C. G., Coleman, S., Jackiw, R.: Ann. Phys. **59**, 42 (1970).
33. Callan Jr., C. G., Gross, D. J.: Phys. Rev. Letters **21**, 311 (1968).
34. Callan Jr., C. G., Gross, D. J.: Phys. Rev. Letters **22**, 156 (1969).

35. Carruthers, P.: Phys. Reports **1**, 1 (1971).
36. Castell, L.: Nucl. Phys. B**4**, 343 (1967).
37. Castell, L.: Nuovo Cimento **46**A, 1 (1966).
38. Castell, L.: Nucl. Phys. B**5**, 601 (1968).
39. Chen, H. H., et al.: Irvine Preprint UCI-10 (1971).
40. Ciccariello, S., Gatto, R., Sartori, G., Tonin, M.: Phys. Letters **30** B, 546 (1969).
41. Ciccariello, S., Gatto, R., Sartori, G., Tonin, M.: Ann. Phys. **65**, 265 (1971).
42. Ciccariello, S., Sartori, G., Tonin, M.: Nuovo Cimento **63** A, 846 (1969).
43. Coleman, S., Jackiw, R.: MIT Preprint (1970).
44. Coleman, S., Jackiw, R.: Ann. Phys. **67**, 552 (1971).
45. Cornwall, J. M., Jackiw, R.: Phys. Rev. D**4**, 367 (1971).
46. Cornwall, J. M., Norton, R.: Phys. Rev. **173**, 1637 (1968).
47. Cornwall, J. M., Norton, R.: Phys. Rev. **177**, 2584 (1969).
48. Crewter, R.: Phys. Rev. D**3**, 3152 (1971).
49. De Alwis, S. P.: Phys. Letters **36** B, 106 (1971).
50. Dell'Antonio, G. F., Frishman, Y., Zwanziger, D.: To be published.
51. Dirac, P. A. M.: Ann. Math. **37**, 429 (1936).
52. Drell, S. D.: Erice Lectures (1969); Stanford Preprint SLAC-PUB-689 (1969).
53. Drell, S. D.: Rapporteur's talk at the Amsterdam Intern. Conf. on Elementary Particles (1971); Stanford Preprint SLAC-PUB-948 (1971).
54. Drell, S., Levy, D. J., Yan, T. M.: Phys. Rev. Letters **22**, 744 (1969).
55. Drell, S. D., Levy, D. J., Van, T. M.: Phys. Rev. **187**, 2159 (1969).
56. Drell, S. D., Levy, D. J., Yan, T. M.: Phys. Rev. D**1**, 1035 (1970).
57. Drell, S. D., Levy, D. J., Yan, T. M.: Phys. Rev. D**1**, 1617 (1970).
58. Drell, S. D., Yan, T. M.: Phys. Rev. D**1**, 2402 (1970).
59. Drell, S. D., Yan, T. M.: Phys. Rev. Letters **24**, 181 (1970).
60. Drell, S. D., Yan, T. M.: Ann. Phys. **66**, 578 (1971).
61. Ellis, J., Weisz, P., Zumino, B.: Phys. Letters **34** B, 91 (1971).
62. Esteve, A., Sona, P. G.: Nuovo Cimento **32**, 473 (1964).
63. Ferrara, S., Gatto, R., Grillo, A. F.: Nucl. Phys. B**34**, 349 (1971).
64. Ferrara, S., Gatto, R., Grillo, A. F.: Phys. Letters **36** B, 124 (1971) and erratum, ibidem.
65. Ferrara, S., Gatto, R., Grillo, A. F.: Phys. Rev. **D5**, 3102 (1972).
66. Ferrara, S., Gatto, R., Grillo, A. F.: Lett. Nuovo Cimento **2**, 1363 (1971).
67. Ferrara, S., Gatto, R., Grillo, A. F.: Frascati Preprint LNF 71—83 (1971), to appear on Annals of Physics.
68. Feynman, R. P.: In III Topical Conf. High Energy Collisions of Hadrons. New York: Stony Brook 1969, and unpublished.
69. Feynman, R. P.: Phys. Rev. Letters **23**, 1415 (1969).
70. Flato, M., Sternheimer, D.: Compt. Rend. **263**, 935 (1966).
71. Frishman, Y.: Phys. Rev. Letters **25**, 960 (1970).
72. Frishman, Y.: Ann. Phys. **66**, 373 (1971).
73. Fritzsch, H., Gell-Mann, M.: Talk presented at the Coral Gables Conf. on Fundamental Interactions at High Energies (1971); Caltech Preprint, CALT-66-297 (1971).
74. Fulton, T., Rohrlich, R., Witten, L.: Nuovo Cimento **26**, 652 (1962).
75. Fulton, T., Rohrlich, R., Witten, L.: Rev. Mod. Phys. **34**, 442 (1962).
76. Gatto, R., Sartori, G.: Nuovo Cimento **7 A**, **99** (1972).
77. Gelfand, I. M., Graev, M. I.: Irv. Akad. Nauk. SSSR, Ser. Math. **29**, 1329 (1965), Intern. Spring School for Theoretical Physics, Yalta (1966).
78. Gelfand, I. M., Ponomarev, V. A.: Russian Math. Surveys.
79. Gell-Mann, M.: Lectures at the Summer School of Theoretical Physics, University of Mawan (1969) Caltech Preprint CALT-68-244 (1969).

80. Gell-Mann, M.: Invited talk at the Coral Gables Conference (1970).
81. Georgelin, Y., Jersak, J., Stern, J.: Orsay Preprint IPNO/TH 188 (1970).
82. Georgelin, Y., Jersak, J., Stern, J.: Orsay Preprint IPNO/TH 198 (1970).
83. Gilman, F. J.: Proc. 4th. Intern. Symp. on Electron and Proton Interaction at High Energies, Liverpool (1969).
84. Gross, L.: J. Math. Phys. **5**, 687 (1964).
85. Gross, D. J., Jackiw, R.: Phys. Rev. **163**, 1688 (1967).
86. Gross, D. J., Jackiw, R.: Phys. Rev. **180**, 1359 (1969).
87. Gross, D. J., Treiman, S. B.: Princeton University preprint (1971).
88. Gross, D. J., Wess, J.: Phys. Rev. D **2**, 753 (1970).
89. Gursey, F.: Nuovo Cimento **3**, 988 (1956).
90. Gursey, F.: Ann. Phys. **24**, 211 (1963).
91. Haberler, P. L. F.: Max-Planck-Institut preprint (1971).
92. Hepner, W. A.: Nuovo Cimento **26**, 352 (1962).
93. Hermann, R.: Lie groups for physicists, Chap. 9. : Benjamin 1966.
94. Ingraham, R. L.: Nuovo Cimento **12**, 825 (1954).
95. Ioffe, B. L.: Phys. Letters **30** B, 123 (1969).
96. Isham, C. J., Salam, A., Strathdee, J.: Phys. Rev. D **2**, 685 (1970).
97. Isham, C. J., Salam, A., Strathdee, J.: Phys. Letters **31** B, 300 (1970).
98. Isham, C. J., Salam, A., Strathdee, J.: Ann. Phys. **72**, 98 (1971).
99. Jackiw, R.: Phys. Rev. **175**, 2058 (1968).
100. Jackiw, R., Van Royen, R., West, G.: Phys. Rev. D **2**, 2473 (1970).
101. Jacobson, M.: Lie Algebras. : Interscience 1962.
102. Jersak, J., Stern, J.: Nuovo Cimento **59**, 315 (1969).
103. Johnson, K.: Nuovo Cimento **20**, 773 (1960).
104. Kastrup, H. A.: Ann. Physik **7**, 388 (1962).
105. Kastrup, H. A.: Nuclear Phys. **58**, 561 (1964).
106. Kastrup, H. A.: Phys. Rev. **142**, 1060 (1966).
107. Kastrup, H. A.: Phys. Rev. **143**, 1041 (1966).
108. Kastrup, H. A.: Phys. Rev. **147**, 1130 (1966).
109. Kastrup, H. A.: Phys. Rev. **150**, 1189 (1966).
110. Kastrup, H. A., et al.: To be published.
111. Kihlberg, A., Muller, V. F., Halbwachs, F.: Commun. Math. Phys. **3**, 194 (1966).
112. Klauder, J. R., Leutwyler, H., Streit, L.: Nuovo Cimento **66**, 536 (1970).
113. Leutwyler, H.: Proc. Summer School for Theoretical Physics, Karlsruhe. Springer Tracts Mod. Phys. **50**, 29 (1966).
114. Leutwyler, H., Stern, J.: Phys. Letters **31** B, 458 (1970).
115. Leutwyler, H., Stern, J.: Nucl. Phys. B **20**, 77 (1970).
116. Lewellyn Smith, C. H.: Nucl. Phys. B **17**, 270 (1970).
117. Lowenstein, J.: Commun. Math. Phys. **16**, 265 (1970).
118. Mack, G.: Nucl. Phys. B **5**, 499 (1968).
119. Mack, G.: Phys. Letters **26** B, 515 (1968).
120. Mack, G.: Phys. Rev. Letters **25**, 400 (1970).
121. Mack, G.: Trieste Preprint IC/70/95 (1970).
122. Mack, G., Salam, A.: Ann. Phys. **53**, 174 (1969).
123. Mack, G., Todorov, J.: J. Math. Phys. **10**, 2078 (1969).
124. Mackey, G. W.: Bull. Ann. Math. Soc. **69**, 628 (1963).
125. Marx, G.: Proc. 1969 Budapest Cosmic Ray Conference.
126. McLennan, J. A.: Nuovo Cimento **3**, 1360 (1956).
127. McLennan, J. A.: Nuovo Cimento **5**, 640 (1957).
128. Migdal, A. A.: Phys. Lett. **37 B**, 98 (1971).
129. Migdal, A. A.: Phys. Lett. **37 B**, 386 (1971).

130. Miller, G., et al.: XVth. Intern. Conf. on High Energy Phys., Kiev (1970).
131. Møller, C.: Ann. Inst. Poincaré, **11**, 251 (1949).
132. Muray, Y.: Progr. Theor. Phys. **9**, 147 (1953).
133. Muray, Y.: Nucl. Phys. **6**, 489 (1958).
134. Myatt, G., Perkins, D. H.: Phys. Letters **34** B, 542 (1971).
135. O'Raifortaigh, L.: Phys. Rev. Letters **14**, 575 (1965).
136. Paschos, E.: Rockefeller, University Preprint.
137. Pokinghorne, J. C.: University of Cambridge Preprint DAMTP-71/20 (1971).
138. Polyakov, A. M.: Sov. Phys.-JETP **28**, 533 (1969).
139. Raczka, R., Limic, M., Niederle, J.: J. Math. Phys. **7**, 1861 (1966).
140. Raczka, R., Limic, M., Niederle, J.: J. Math. Phys. **8**, 1079 (1967).
141. Rohrlich, F.: Lectures at the Intern. Universitätswochen für Kernphysik, Schladming (1971).
142. Salam, A., Strathdee, J.: Phys. Rev. **184**, 1750 (1969).
143. Salam, A., Strathdee, J.: Phys. Rev. **184**, 1760 (1969).
144. Sartori, G.: Lettere Nuovo Cimento **4**, 583 (1970).
145. Schreier, E. J.: Phys. Rev. D **3**, 980 (1971).
146. Schwinger, J.: Phys. Rev. **130**, 406 (1963).
147. Suri, A., Yennie: Stanford University Preprint.
148. Symanzik, K.: Commun. Math. Phys. **16**, 49 (1970).
149. Symanzik, K.: Commun. Math. Phys. **18**, 227 (1970).
150. Symanzik, K.: Coral Gables Conf. on Fundamental Interactions at High Energies. Ed. by A. Palmuther, G. J. Iverson and R. M. Williams. : Gordon and Breach 1970.
151. Symanzik, K.: Springer Tracts Mod. Phys. **57**, 222 (1971).
152. Takhelidze, A. N.: 1970 Coral Gables Conference.
153. Taylor, R.: Proc. 4th. Intern. Symp. on Electron and Proton Interactions at High Energies, Liverpool (1969).
154. Thirring, W.: Ann. Phys. **3**, 91 (1958).
155. Tonin, M.: Nuovo Cimento **47** A, 919 (1967).
156. Yao, T.: J. Math. Phys. **8**, 1931 (1967); **9**, 615 (1968).
157. Weisskopf, V. F.: Invited talk at the Topical Seminar on Electromagnetic Interactions, Trieste (1971).
158. Wess, J. E.: Nuovo Cimento **18**, 1086 (1960).
159. Wilson, K.: Phys. Rev. **179**, 1499 (1969).
160. Wilson, K.: Phys. Rev. D **2**, 1473 (1970).
161. Wilson, K.: Stanford University Preprint, SLAC-PUB-737 (1970).
162. Wilson, K.: Rapporteur's talk at the 1971 Cornell Conference on Electromagnetic Interactions, Ed. N. B. Mistry (Cornell University 1971).
163. Zimmermann, W.: Commun. Math. Phys. **6**, 161 (1967).
164. Zimmermann, W.: Commun. Math. Phys. **8**, 66 (1968).
165. Zimmermann, W.: 1970 Brandeis Summer Institute in Theoretical Phys., Vol. 1. Ed. by S. Desai, M. Grisaru, H. Pardleton, MIT Press 1971.
166. Zumino, B.: 1970 Brandeis Summer Institute in Theoretical Phys., Vol. II. Ed. by S. Desai, M., Grisaru, H. Pendleton, MIT Press 1971.
167. Zumino, B.: Rapporteur's talk at the 1970 Kiev Conference.

Received: February 1972

Dr. S. Ferrara and Dr. A. F. Grillo
Laboratori Nazionali di Frascati del CNEN, Frascati, Italy

Prof. Dr. R. Gatto
Instituto di Fisica dell'Università, Roma
Istituto Nazionale di Fisica Nucleare, Sezione di Roma, Italy

SPRINGER TRACTS IN MODERN PHYSICS

Ergebnisse der exakten Naturwissenschaften

Atomic Physics

Dettmann, K.: High Energy Treatment of Atomic Collisions (Vol. 58)
Donner, W., Süßmann, G.: Paramagnetische Felder am Kernort (Vol. 37)
Racah, G.: Group Theory and Spectroscopy (Vol. 37)
Seiwert, R.: Unelastische Stöße zwischen angeregten und unangeregten Atomen (Vol. 47)
Zu Putlitz, G.: Determination of Nuclear Moments with Optical Double Resonance (Vol. 37)

Elementary Particle Physics

Current Algebra

Furlan, G., Paver, N., Verzegnassi, C.: Low Energy Theorems and Photo- and Electroproduction Near Threshold by Current Algebra (Vol. 62)
Gatto, R.: Cabibbo Angle and $SU_2 \times SU_2$ Breaking (Vol. 53)
Genz, H.: Local Properties of σ-Terms: A Review (Vol. 61)
Kleinert, H.: Baryon Current Solving SU (3) Charge-Current Algebra (Vol. 49)
Leutwyler, H.: Current Algebra and Lightlike Charges (Vol. 50)
Mendes, R. V., Ne'eman, Y.: Representations of the Local Current Algebra. A Constructional Approach (Vol. 60)
Müller, V. F.: Introduction to the Lagrangian Method (Vol. 50)
Pietschmann, H.: Introduction to the Method of Current Algebra (Vol. 50)
Pilkuhn, H.: Coupling Constants from PCAC (Vol. 55)
Pilkuhn, H.: S-Matrix Formulation of Current Algebra (Vol. 50)
Renner, B.: Current Algebra and Weak Interactions (Vol. 52)
Renner, B.: On the Problem of the Sigma Terms in Meson-Baryon Scattering. Comments on Recent Literature (Vol. 61)
Soloviev, L. D.: Symmetries and Current Algebras for Electromagnetic Interactions (Vol. 46)
Stech, B.: Nonleptonic Decays and Mass Differences of Hadrons (Vol. 50)
Stichel, P.: Current Algebra in the Framework of General Quantum Field Theory (Vol. 50)
Stichel, P.: Current Algebra and Renormalizable Field Theories (Vol. 50)
Stichel, P.: Introduction to Current Algebra (Vol. 50)
Verzegnassi, C.: Low Energy Photo and Electroproduction, Multipole Analysis by Current Algebra Commutators (Vol. 59)
Weinstein, M.: Chiral Symmetry. An Approach to the Study of the Strong Interactions (Vol. 60)

Electromagnetic Interactions

Deep Inelastic Lepton Scattering

Drees, J.: Deep Inelastic Electron-Nucleon Scattering (Vol. 60)
Landshoff, P. V.: Duality in Deep Inelastic Electroproduction (Vol. 62)
Llewellyn Smith, C. H.: Parton Models of Inelastic Lepton Scattering (Vol. 62)
Rittenberg, V.: Scaling in Deep Inelastic Scattering with Fixed Final States (Vol. 62)
Rubinstein, H. R.: Duality for Real and Virtual Photons (Vol. 62)
Rühl, W.: Application of Harmonic Analysis to Inelastic Electron-Proton Scattering (Vol. 57)

Experimental Techniques

Panofsky, W. K. H.: Experimental Techniques (Vol. 39)

Form Factors, Colliding Beam Experiments

Photo- and Electroproduction of Pions

Quantum Electrodynamics

Field Theory, Light Cone Singularities

Strong Interactions

Mesons and Baryons

Regge Pole Theory, Dual Models

Symmetries

Weak Interactions

Nuclear Physics

Baryon-Baryon-Scattering

Electron Scattering

To Appear in Volume 68

Schmid, D.: Nuclear Magnetic Double Resonance – Principles and Applications in Solid State Physics
Bäuerle, D.: Vibrational Absorption of Electron and Hydrogen Centers in Ionic Crystals
Behringer, J.: Factor Group Analysis Revisited and Unified

To Appear in Forthcoming Volumes:

Börner, G.: On the Properties of Matter in Neutron Stars
Stewart, J., Walker, M.: Black Holes: the Outside Story
Überall, H.: Study of Nuclear Structure by Muon Capture
Levinger, J. S.: Two-Nucleon and Three-Nucleon Systems
Brandmüller, J., Claus, R.: Light Scattering on Optical Phonons and Polaritons
Langbein, D.: Theory of van der Waals Attraction